Evaluation Report of National Marine Innovation Index 2016

The First Institute of Oceanography State Oceanic Administration

China Ocean Press

2019 • Beijing

图书在版编目（CIP）数据

国家海洋创新指数报告. 2016 = Evaluation Report of National Marine Innovation Index 2016 : 英文 / 国家海洋局第一海洋研究所著. — 北京 : 海洋出版社, 2018.11

ISBN 978-7-5210-0143-3

Ⅰ. ①国… Ⅱ. ①国… Ⅲ. ①海洋经济－技术革新－研究报告－中国－2016－英文 Ⅳ. ①P74

中国版本图书馆CIP数据核字(2019)第030927号

责任编辑：苏　勤
责任印制：赵麟苏

海洋出版社 出版发行
http://www.oceanpress.com.cn
北京市海淀区大慧寺路 8 号　　邮编：100081
北京朝阳印刷厂有限责任公司印刷
2019年1月第1版　　2019年1月第1次印刷
开本：889mm × 1194mm　1 / 16　印张：10
字数：300千字　　定价：98.00元
发行部：62132549　　邮购部：68038093　　总编室：62114335
海洋版图书印、装错误可随时退换

Evaluation Report of National Marine Innovation Index 2016

Editorial Committee

Consultants: Ding Dewen, Jin Xianglong, Wu Lixin, Qu Tanzhou, Li Tiegang, Xin Hongmei, Wang Xiaoqiang, Feng Lei, Yu Xingguang, Ma Deyi, Wei Zexun, Wang Zongling, Lei Bo, Zhang Wen, Wen Quan, Wang Baodong, Feng Aiping, Wang Yuan

Editor-in-chief: Liu Dahai, He Guangshun

Editorial members: Gao Feng, Gao Runsheng, Pan Kehou, Xu Xingyong, Li Renjie

Compilers: The First Institute of Oceanography State Oceanic Administration
National Marine Data and Information Service
The Lanzhou Documentation and Information Center of the Chinese Academy of Sciences
Qingdao National Laboratory of Marine Science and Technology

Compilation group members: Liu Dahai, Lu Wenhai, Wang Chunjuan, Xu Meng, Li Xianjie, Wang jinping, Lu Jingliang, Guo Yue, Lin Xianghong, Yin Xigang, Li Dahai, Li Sen, Li Xiaoxuan, An Chenxing

Calculation group members: Liu Dahai, Li Xianjie, Xu Meng, Li Xiaoxuan

Translation group members: Yang Hong, Sun Guangfeng, Zou Shanshan, Xin Haiyan, Yu Ying

Preface

As the top priority of five development concepts, innovation shows the general development trend and marks the direction of comprehensive and in-depth reform of China. The 18th National Congress of the CPC put forward that we must place the implementation of innovation-driven development strategy at the center of national overall development, highlighting that scientific and technological (S&T) innovation is the strategic support for the improvement of social productivity and national comprehensive strength. The "*Outline of National Innovation-driven Development Strategy*" pointed out forward that we must put S&T innovation on top priority to promote all-round innovation, invigorate innovative dynamism with institutional reform, support the building of high-level innovation-oriented country with high-efficiency innovative system, promote the fundamental transformation of economic and social development, and provide a powerful driving force for the realization of the China dream of the great national rejuvenation. The "*Vision and Actions on Jointly Building Silk Road Economic Belt and the 21st-Century Maritime Silk Road*" put forward the development ideas to "create new systems and mechanisms of an open economy, step up S&T innovation, develop new advantages for participating in and leading international cooperation and competition, and become the pace-setter and main force in the '*Belt and Road Initiative*', particularly in the building of the 21st-Century Maritime Silk Road". On July 28, 2016, the State Council issued "*The '13th Five-Year Plan' for National Scientific and Technological Innovation*" and put forward exploring the new realm of development, striding forward to become an innovation-driven country and accelerating the building of a S&T power of the world.

The period of "*the 13th Five-Year*" is the decisive stage of transforming China into a well-off society in an all-round way and the key stage of implementing innovation-driven development strategy and building marine power. Marine innovation is an important component of national innovation, and also the driving source of the strategy for the realization of a marine power. The first working conference on strategic research of the "*Overall Planning of Marine Scientific and Technological Innovation*" proposed that efforts should be made " to conduct the research of marine strategy on the basis of 'overall planning' and 'innovation'", "to recognize the innovation path and modes, and to take stock of the 'overall resources'". On May 8, 2017, Ministry of Science and Technology, Ministry of Land and Resources and State Oceanic Administration jointly issued "*Special Planning of 'the 13th Five year' Scientific and Technological Innovation in Marine Field*", which highlighted the importance of further constructing and improving national S&T innovation system, enhancing national marine S&T innovation capability, significantly strengthening the supporting role that S&T innovation plays in the development of marine industry.

In response to the national marine innovation strategy and for the sake of building of a national innovation system, the First Institute of Oceanography, State Oceanic Administration (SOA) set about carrying out the measurement and calculation of marine innovation indicators as from 2006, and initiated the research on national marine innovation indexes in 2013. With the help and support of the leaders and scholars from SOA, the first issue of series of evaluation reports on national marine innovation index "*Tentative Evaluation Report of National Marine Innovation Index 2013*" was officially published in 2015. The second issue "*Tentative Evaluation Report of National Marine Innovation Index 2014*" and the third issue "*Evaluation Report of National Marine Innovation Index 2015*" were published in December 2015 and December 2016 respectively. The "*Evaluation Report of National Marine Innovation Index 2016*" is the fourth issue of the series of evaluation reports.

The "*Evaluation Report of National Marine Innovation Index 2016*" follows the evaluation methods of national marine innovation indexes used in the previous series of evaluation reports and establishes the index system for national marine innovation indexes in terms of marine innovation resources, marine knowledge creation, marine enterprise innovation, marine innovation performance and marine innovation environment on the basis of data such as marine economic statistics, science and technology statistics, and technological achievements registration. It has quantitatively measured and assessed national marine innovation indexes from 2001 to 2015, made an objective analysis of both state and regional marine innovation capabilities and conducted special analysis for the regional demonstration of marine economic innovation development of our country, output efficiency of national marine S&T input and international marine S&T research situations, effectively reflecting the quality and efficiency of marine innovation in China.

Commissioned by the Department of Science and Technology, Marine Policy Research Center of the First Institute of Oceanography of SOA organized the compilation of "*Evaluation Report of National Marine Innovation Index 2016*". Lanzhou Documentation and Information Center of the Chinese Academy of Sciences co-wrote the marine papers and patents, analyzed the situations of international marine science and technology research and other parts. Qingdao National Laboratory of Marine Science and Technology did special analysis to marine national laboratories. The Department of Science and Technology of SOA provided the relevant content of the regional demonstration of marine economic innovation and development in China. National Marine Data and Information Service, Innovation and Development Department of the Ministry of Science and Technology, Science and Technology Department of the Ministry of Education, School of Management, Huazhong (Central China) University of Science and Technology and other units and departments provided data support. Here sincere thanks go to the Department of Science and Technology, SOA, as well as units and individuals who contributed to its compilation and data support.

We hope that this series of evaluation reports on national marine innovation index can act as a window to the whole society to gain an understanding of marine innovation development in China. Since this report is a staged attempt in the evaluation studies of national marine innovation index, we earnestly accept criticisms and suggestions from our colleagues if there are problems or deficiencies. Our compilation group will absorb the valuable opinions of experts and scholars in order to keep improving the series of evaluation reports on national marine innovation index. Please send feedback to mpc@fio.org.cn.

The First Institute of Oceanography State Oceanic Administration

August, 2017

Contents

I. Introduction

The strategic objective of building a "maritime power" was formally incorporated into the national grand strategy in 2012 at the 18th National Congress of the CPC. Realizing the goal of being a "maritime power" is a key part of the "China dream" for the great rejuvenation of the Chinese nation and marine innovation is an indispensable component in the process of achieving it. The "*13th Five-Year Plan*" is a key period for achieving strategic breakthroughs in marine science and technology, during which marine innovation is desperately needed for the development of marine economy.

"*Evaluation Report of National Marine Innovation Index 2016*" objectively analyzes the current status and development trends of marine innovation in China, establishes China's national marine innovation index, quantitatively assesses national and regional marine innovation capabilities, makes evaluation and prospect on marine innovation capabilities of China, and conducts specific analysis of the key issues of marine innovation about international marine science and technology research situations. Specifically, it is divided into seven parts:

Chapter I. *Introduction*. A comprehensive account of the significance of marine innovation is given and an overall introduction to the "*Evaluation Report of National Marine Innovation Index 2016*" is made in this chapter.

Chapter II. *Marine innovation in China viewed from the perspective of data*. A comprehensive analysis of the current status of the development of marine innovation in China is made through six main indicators: human resources for marine innovation, national platform of marine innovation, funding scale for marine innovation, marine innovation output achievements, marine innovation activities at higher institutions and marine innovation knowledge services.

Chapter III. *Evaluation and analysis of national marine innovation index*. Quantitative evaluation of national marine innovation index in China from 2001-2016 is conducted. The conclusion shows that national marine innovation index in China witnesses a significant increase with an average annual growth rate of 21.90%, among which the sub-index of marine innovation resources maintains an upward trend with an average annual growth rate of 6.92%; the sub-index of marine knowledge creation witnesses a robust growth with an average annual growth rate of 21.80%, in line with the growth rate of national marine innovation index; the sub-index of marine enterprise innovation has a sharp annual increasing rate of 64.23%, which is the fastest increasing rate of the five sub-indexes; the sub-index of marine innovation performance presents relatively slower growth trend among the five sub-indexes with an average annual growth rate of 4.54%; the sub-index of marine innovation environment keeps an upward trend with an average annual growth rate of 10.78%.

Chapter IV. *Evaluation and analysis of regional marine innovation index*. Quantitative assessment of regional marine innovation index of 2015 in China is conducted. The conclusion shows that judging from the coastal provinces (municipalities) of China, Shanghai has the highest score in regional marine innovation index, closely followed by Shandong, Guangdong and Tianjin; in terms of the five economic zones, the Pearl River Delta Economic Zone has the highest score in regional marine innovation index, followed by the Yangtze River Delta Economic Zone, Bohai Rim Economic Zone, West Coast of the Economic Zone and the Beibu Gulf Rim Economic Zone; from the viewpoint of three marine economic circles, China's marine economic circles are characterized by the stronger northern and eastern circles and weaker southern circle.

Chapter V. *Progress and prospect of China's marine innovation capability*. Based on the assessment results and situation analysis in the above chapters, we have conducted comprehensive evaluation of China's marine innovation capability and development status, and made forecast about the development of China's marine innovation.

Chapter VI. *Specific analysis on regional demonstration of China's marine economic innovation development*. We have summarized the implementation situations of regional demonstration of China's marine economic innovation development of 2015 and analyzed their construction effects.

Chapter VII. Specific Analysis of the Input-output Efficiency of China's Marine Science and Technology. With city as basic research unit, marine science and technology input-output efficiency as research subject, we use the DEA model to calculate the input-output efficiency of marine-related cities of China and conduct retrospective analysis and trend prediction to the input-output efficiency of marine science and technology from the *"10th Five-Year"* period to the *"13th Five-Year"* period.

II. Marine Innovation in China Viewed from the Perspective of Data

Under the background of implementing the strategy of maritime power, China has continuously made new major achievements in the field of marine S&T innovation development, significantly boosted its independent innovation capability, and markedly enhanced S&T competitiveness and overall strength. In some areas, China has reached internationally of advanced level, and the number of the national award-winning S&T achievements, academic papers and patents has been significantly improved, and the conditions and environment for marine innovation have been enhanced dramatically.

This report analyzes the development situations of marine innovation in China in terms of six main indicators: human resources for marine innovation, national platform for marine innovation, funding scale for marine innovation, marine innovation achievements, marine innovation activities of higher institutions, and marine innovation knowledge services.

Human resources for marine innovation are constantly optimized. The structure of staff involved in S&T activities at marine research institutes is continuously improved. Both of the overall number of research and development (R&D) staff and the workload equivalent to full-time work rise steadily. The structure of academic qualifications of R&D personnel is further optimized, and the workload equivalent to full-time work of R&D staff is reasonably constituted.

The number of national platforms for marine innovation has been gradually increased. The number of national (key/engineering) laboratories of marine scientific research institutes and the number of national engineering (research/technology research) centers have been significantly increased. The capital construction and the fixed assets of marine scientific research institutes have been increasing year by year.

The funding scale for marine innovation has significantly boosted. The R&D funding scale for marine scientific research institute has significantly improved and the internal spending of R&D funding has grown steadily.

Marine innovation achievements have grown steadily. The total number of marine S&T papers from marine scientific research institutes maintains its momentum and the number of papers published and recorded on the Science Citation Index (SCI) in the marine field has grown substantially, with significantly improved citations. The types of publications on marine science and technology have obviously increased, the number of patent applications and grants shows robust growth, and the revenues from the transfer and licensing of ownership of invention patents have gradually increased.

Marine innovation at the higher institutions is well developed. Marine-related higher institutions are witnessing year-by-year growth in personnel, funding, projects, and so on.

Marine science and technology have made gradual, steady contribution to the development of the marine economy. The contribution rate of marine S&T progress in 2015 was 64.2%[1]. The transformation rate of marine S&T achievements reached 50.4%[2], and marine S&T innovation plays an increasingly prominent role in promoting achievement transformation.

1 The contribution rate of marine S&T progress in 2015 refers to the average value measured based on the relevant data from 2006 to 2015.

2 The transformation rate of marine S&T achievements in 2015 is obtained based on the measurement of relevant data from 2000 to 2015.

1. Stable Structure of Human resources for Marine Innovation

Human resources for marine innovation are the main force and strategic resources for the building of a maritime power and an innovation-driven country. The overall quality of marine innovation researchers determines the speed and range in boosting national marine innovation capability. The personnel involved in S&T activities and R&D at marine research institutes are important human resources for marine innovation, reflecting the talent reserve status of human resources for national marine innovation. Among them, the personnel involved in S&T activities refers to the personnel engaged in S&T activities at the marine scientific research institutes, including technological management staff, personnel involved in project activities, and technical service personnel; R&D personnel refers to the staff of the marine scientific research institutes and external researchers, and the postgraduate students who participate in R&D projects, R&D project management staff, and the staff providing direct services for R&D activities.

1.1 The structure of staff involved in S&T activities has been continuously optimized.

With regard to personnel structure, from 2011 to 2015, the proportion of staff engaged in project activities at marine scientific research institutes of China (namely the staff of the research office or research group) in staff involved in S&T activities has maintained above 65%, with 2015 witnessing slight drop, while the S&T management personnel (namely the institute heads, business and management personnel) and technical service personnel (namely all types of personnel working directly for S&T services) both have accounted for less than 15%, with 2012 and 2015 witnessing a relatively higher percentage (See Figure 2-1). Judging from the academic qualifications of staff, over the past five years, the proportion of doctoral and master graduates among the staff of China's marine scientific research institutes engaged in S&T activities has shown an overall upward trend. In 2015, the doctoral and master graduates accounted for 23.35% and 32.14% respectively of the total personnel involved in S&T activities, both witnessing an increase compared to 2014 (See Figure 2-2). Judging from the professional title structure of the staff, over the past five years, the proportion of the S&T staff holding senior and intermediate professional titles in Chinese marine scientific research institutes has been twice that of staff with junior professional titles. In 2015, the staff with senior and intermediate professional titles accounted for 39.63% and 33.31% of the total S&T staff (See Figure 2-3).

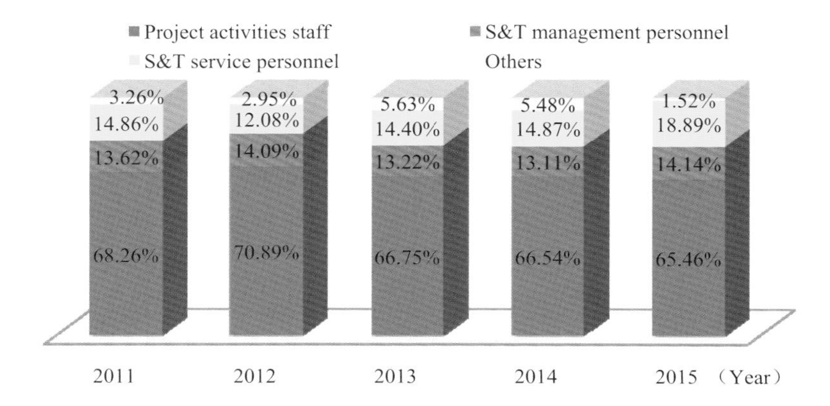

Figure 2-1 Composition of S&T Personnel at Marine Scientific Research Institutes from 2011 to 2015

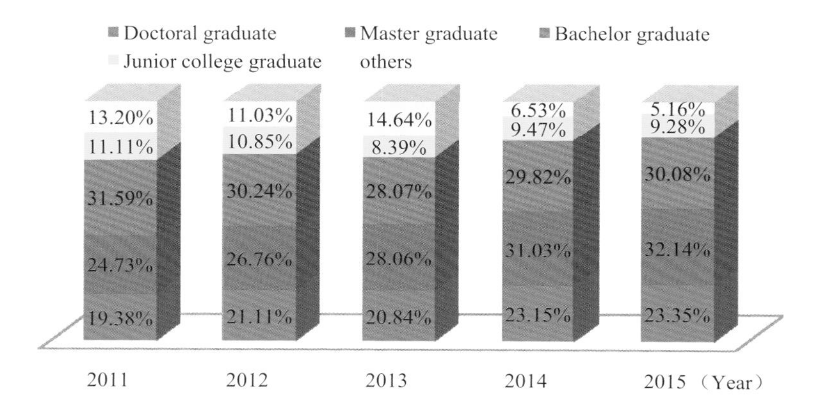

Figure 2-2 Academic Qualifications Structure of S&T Personnel at Marine Scientific Research Institutes from 2011 to 2015

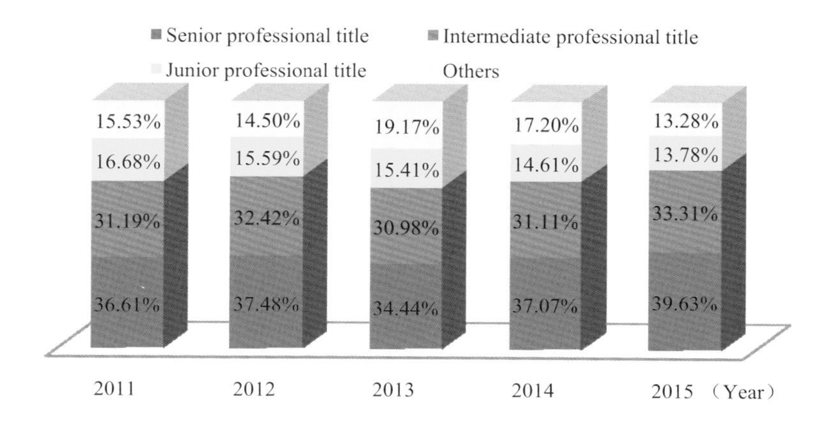

Figure 2-3 Professional Title Structure of S&T Personnel at Marine Scientific Research Institutes from 2011 to 2015

1.2 The total number of R&D personnel and the workload equivalent to full-time work increase steadily.

The total number of R&D personnel and the workload equivalent to full-time work at marine scientific research institutes of China present steady increase trend (See Figure 2-4). From 2001 to 2006, the increases had been relatively slower; from 2006 to 2007, they presented sharp increase, with growth rate at 115.91% and 88.25% respectively; from 2008 to 2009, they presented another rapid increase, with growth rate at 49.62% and 55.18% respectively; from 2009 to 2014, they had resumed a stead increase trend. Both decreased slightly in 2015 compared to 2014.

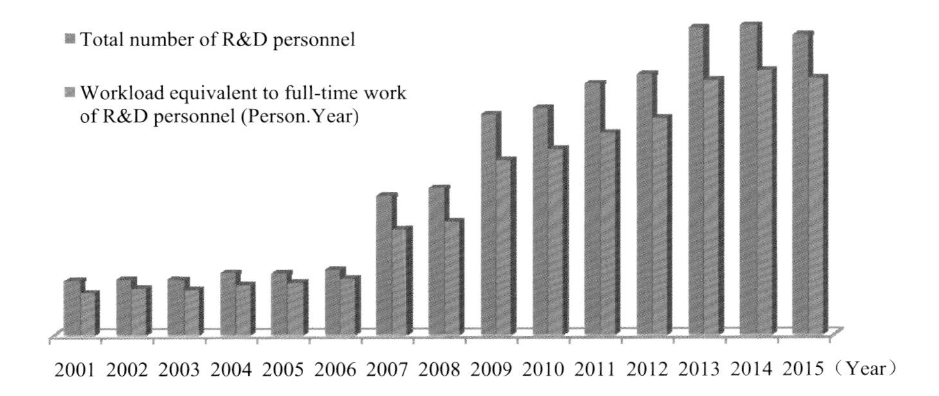

Figure 2-4 Total Number of R&D Personnel at Marine Scientific Research Institutes and the Workload Equivalent to Full-time Work from 2001 to 2015

1.3 The structure of academic qualifications of R&D personnel is further optimized.

In the past 5 years, the number of doctoral graduates in R&D staff of Chinese marine scientific research institutes has kept an upward trend, but the proportion has shown an upward trend in fluctuation. Both the number and proportion of master graduates have witnessed a continuous increase. In 2015, graduates with doctoral and master's degrees accounted for 29.47% and 34.04% of the total number of R&D personnel respectively (See Figure 2-5). The number of doctoral graduates accounted for the highest proportion in 2015, reaching 29.47%, an increase of 3.49% compared to 2011. The proportion of master graduates has maintained steady growth for five consecutive years, up by 6.40% in 2015 over the year 2011.

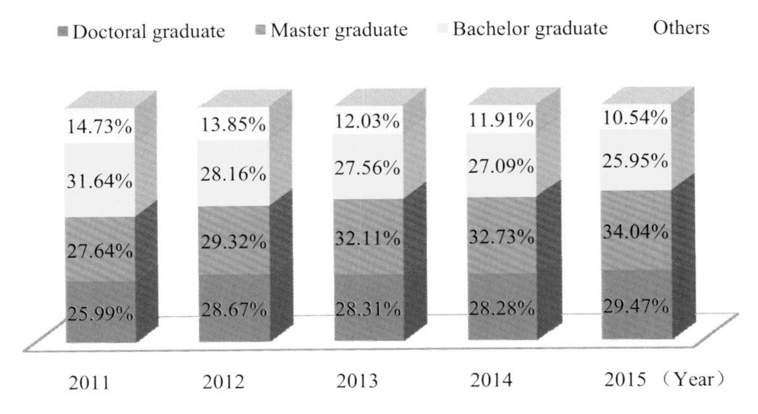

Figure 2-5 Academic Qualification Structure of R&D Personnel at Marine Scientific Research Institutes from 2011 to 2015

1.4 The structure of workload equivalent to full-time work of R&D staff is reasonable.

The workload equivalent to full-time work of R&D staff consists of the workload equivalent to full-time work by researchers, technicians and other support staff. Of whom, researchers refer to the professionals engaged in the conception or creation of new knowledge, new products, new technologies, new methods, and new systems, and the senior management staff of R&D projects as well; technicians refer to those who participate in R&D projects under the guidance of the researchers and use the relevant principles and methods to perform R&D tasks, such as literature retrieval, computer programming, etc; other support staff refer to the secretaries, clerical staff, and administrative personnel, etc. participating in R&D projects or directly assisting in these projects. Over the past five years, the structure of the workload equivalent to full-time work of the R&D staff of marine scientific research institutes in China has been basically stable, showing a reasonable development trend with the workload of researchers accounting for over 60%. In 2015, the workload equivalent to full-time work of the researchers accounted for 60.51% (See Figure 2-6).

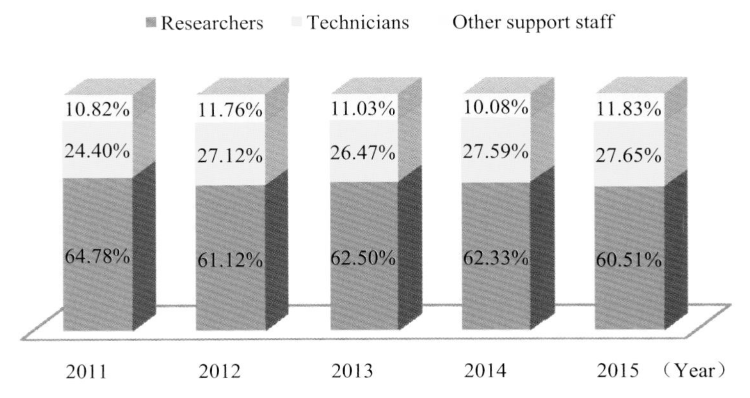

Figure 2-6 Workload Equivalent to Full-time Work by the R&D Staff at Marine Scientific Research Institutes from 2011 to 2015

2. Gradual Improvement of Marine Innovation Platform Environment

2.1 The number of national (key/engineering) laboratories and national engineering (research / technology research) centers of marine research institutes has increased.

From 2002 to 2015, the number of national (key/engineering) laboratories and national engineering (research / technology research) centers has shown an overall increasing trends. And the number of national (key/engineering) laboratories reached the maximum value in 2010. The number of national engineering centers has maintained the maximum value between 2013 and 2015 (See Figure2-7).

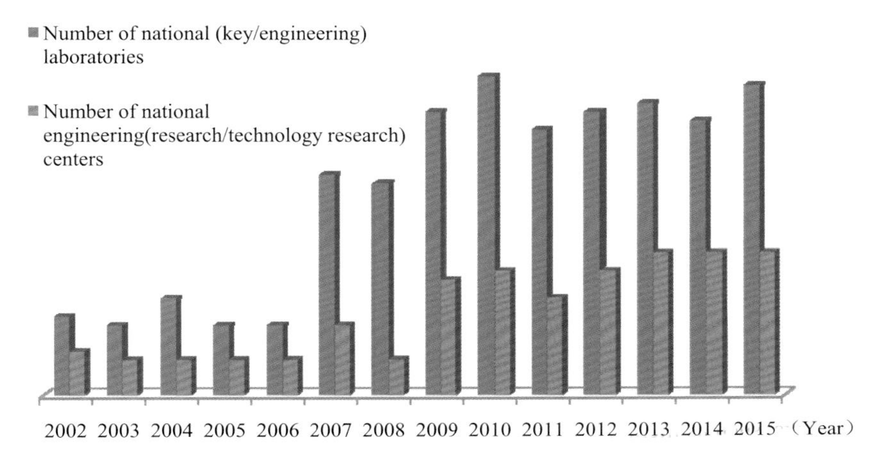

Figure 2-7 Number of National (key/engineering) Laboratories and Engineering (research / technology research) Centers of Marine Scientific Research Institutes from 2002 to 2015

2.2 The actual completion of the capital construction investment has shown steady increase.

The actual completion of capital construction investment refers to the capital construction workload completed by the institutes in the current year, which is expressed in currency. It is divided into the following parts by purposes, namely, instrument and equipment for scientific research, civil engineering for scientific research, civil engineering and equipment for production and operation as well as civil engineering and equipment for living. Capital construction investment in scientific research equipment refers to the total value of purchased instrument and equipment for scientific research in the actual completion of capital construction investment. Capital construction investment in civil engineering for scientific research refers to the completed workload of civil engineering for scientific research in the actually completed amount of capital construction investment (such as scientific research building, laboratory building, etc.). From 2001 to 2015, the actual completion of capital construction investment of China's marine scientific research institutes has shown a steady increase (See Figure 2-8) with the most rapid growth appearing in 2009 and an annual growth rate reaching 307.77%. The growth rate of 2015 was 32 times that of 2001. In terms of purposes, the actual completion of capital construction investment from 2001 to 2015 has been mainly used for civil engineering, instrument and equipment for scientific research (See Figure 2-9). The proportions of these two items in 2015 were respectively 67.11% and 31.84%.

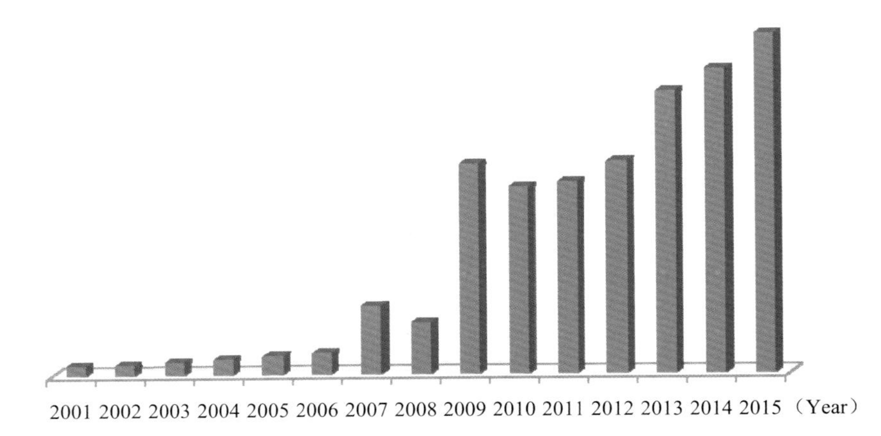

Figure 2-8 The Actual Completion of Capital Construction Investment of Marine Scientific Research Institutes from 2001 to 2015 (thousand Yuan)

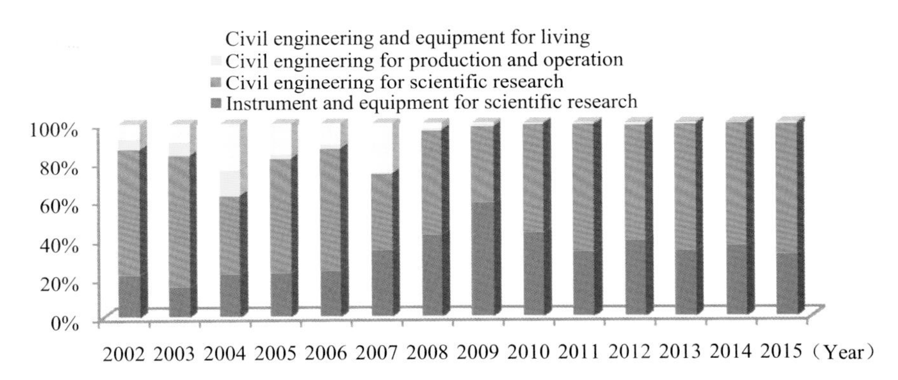

Figure 2-9 The Actual Completion of Capital Construction Investment of Marine Scientific Research Institutes from 2001 to 2015

2.3 Fixed assets and scientific instruments and equipment increase year by year.

Fixed assets refer to the facilities and equipment which can be used for a long time. Although their value can be consumed, their original physical form can be maintained, such as houses and buildings. Properties and materials as fixed assets should meet two requirements at the same time: a durable period of more than one year and a unit value higher than the regulated standard. From 2001 to 2015, the original price of fixed assets of marine scientific research institutes in China has maintained growth trend (See Figure 2-10), with an average annual growth rate of 28.88%. Scientific instruments and equipment in fixed assets refer to the instruments and equipment for scientific research directly used by personnel engaged in S&T activities excluding all kinds of power equipment, machinery and equipment, auxiliary equipment which support capital construction, general transportation equipment (other than means of transportation for scientific investigation), instruments and equipment designated for production. From 2001 to 2015, the original price of the scientific instruments and equipment in fixed assets of China's marine scientific research institutes also has maintained a growth trend (See Figure 2-10), with an average annual growth rate of 34.64%.

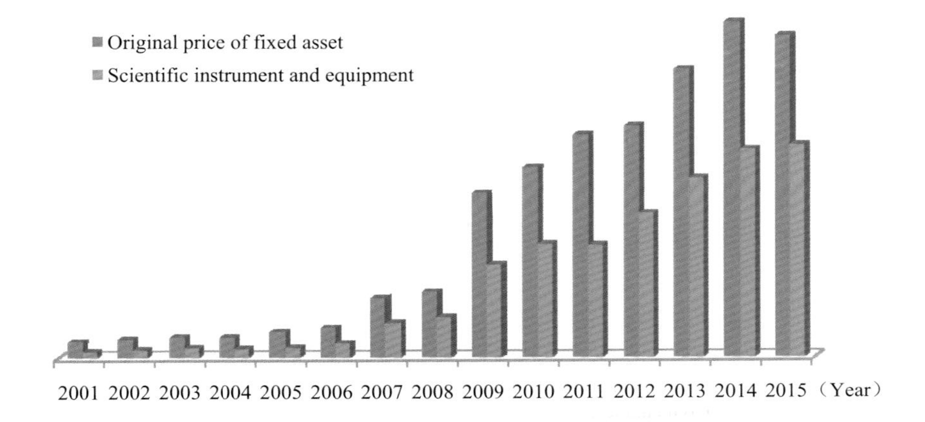

Figure 2-10 Original Prices of Fixed Assets (thousand Yuan) and the Scientific Instruments and Equipment (thousand Yuan) of Marine Scientific Research Institutes from 2001 to 2015

3. Significant Increase in Funding Scale for Marine Innovation

R&D activities are the core part of innovation activities. Not only are they a source of knowledge creation and independent innovation capability, but also the basis of ability to absorb in new knowledge and create new technologies in the context of globalization. To a greater extent, they are important indicators reflecting the coordinated development among economy, science and technology, as well as measuring the quality of economic growth. R&D funds for marine research institutes is important for marine innovation, which can effectively reflect the scale of national marine innovation activities and objectively evaluate national marine S&T strength and innovation capability.

3.1 R&D funding scale grows rapidly.

Since the 21st century, R&D expenditures in China's marine scientific research institutes have been maintaining a growth trend. The year 2009 witnessed the most rapid growth in this indicator, with annual growth rate of 106.46%. From 2001 to 2015, the average annual growth rate has reached 32.16%. The proportion of R&D funds in the national gross ocean product (GOP) generally acts as the indicator revealing the investment degree of national marine scientific research funds, and reflecting the input strength of national marine innovation funds. From 2001 to 2015, this indicator has shown an overall growth trend, with an average annual growth rate of 15.34% and a slight decrease in 2015 compared to 2014 (See Figure 2-11).

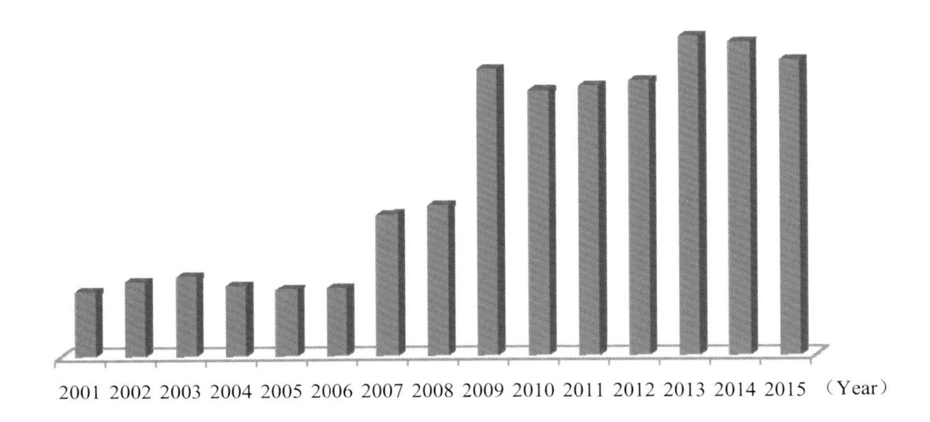

Figure 2-11 Proportion of R&D Funds in National Gross Ocean Product from 2001 to 2015

3.2 The internal spending of R&D funding witnesses steady growth.

The internal spending of R&D funding refers to all expenditures incurring in the current year for carrying out R&D activities by the institutes, including R&D regular expenses and R&D capital construction expenses. From 2001 to 2015, the proportion of R&D capital construction expenses in the internal spending of R&D funding has gradually increased from 5.28% in 2001 to 16.35% in 2015, reflecting the greater emphasis placed on capital construction investment in China (See Figure 2-12).

In terms of the category of expenses, R&D regular expenses include personnel costs (including salary), equipment procurement costs, other daily expenses (including operating expenses and overheads), and R&D capital construction expenses include instruments, equipment costs and civil engineering expenses. From 2001 to 2015, other daily expenses have accounted for over 50% of R&D regular expenses, and the proportion of staff costs and equipment procurement costs has witnessed a small decline (See Figure 2-13); in 2015, staff costs and equipment procurement costs accounted for 31.23 % and 12.27% of the R&D regular expenses respectively and other daily expenses took up 56.51%; from 2001 to 2015, the structure of R&D capital construction expenditures has shown fluctuation trend, among which, the proportion of civil engineering costs has exceeded that of equipment and instruments with the year of 2007 and 2009 as exception when the proportion of civil engineering costs was less than that of equipment and instruments (See Figure 2-14).

In terms of the category of activities, from 2001 to 2015 the proportion of R&D regular expenses used for basic research has witnessed little change (See Figure 2-15). The proportion of funds used for applied research fell from 47.17% in 2001 to 37.41% in 2015, and the proportion of funds used for experimental development rose from 29.26% in 2001 to 36.86% in 2015.

In terms of the sources of funds between 2001 and 2015, the internal spending of R&D funding came mainly from government funds and corporate funds. In addition, the proportion of government funds has declined gradually (See Figure 2-16), while that of corporate funds has been on the increase. In 2015, the proportions of government funds and corporate funds took 71.03% and 18.034% respectively.

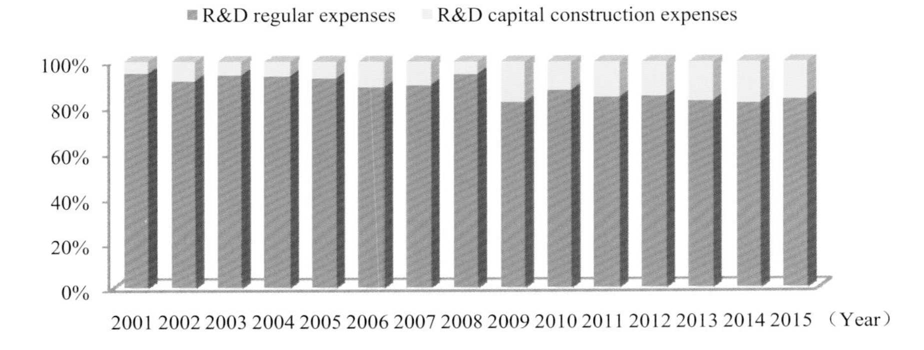

Figure 2-12 Internal Spending of R&D Funds from 2001 to 2015

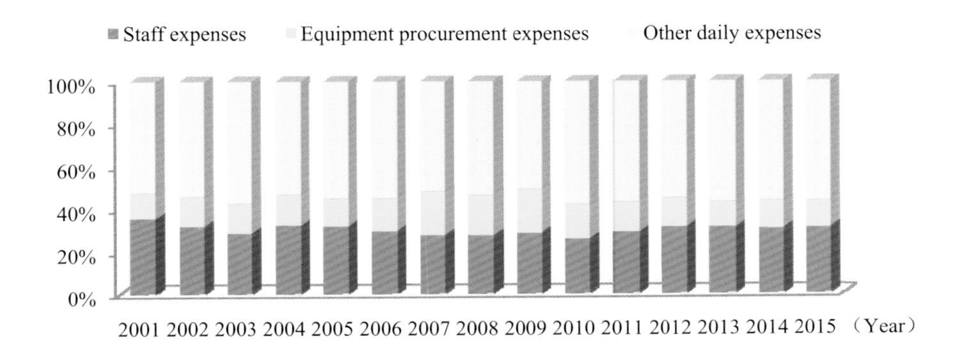

Figure 2-13 Expenditure of R&D Regular Expenses (by Cost Categories) from 2001 to 2015

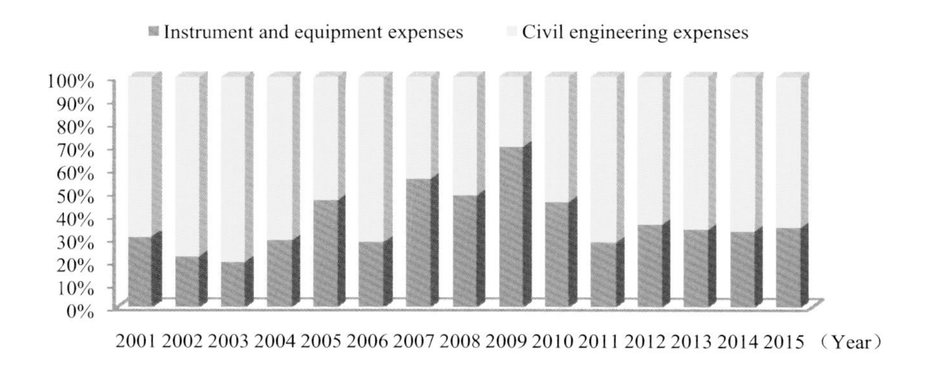

Figure 2-14 R&D Capital Construction Expenses (by Cost Categories) from 2001 to 2015

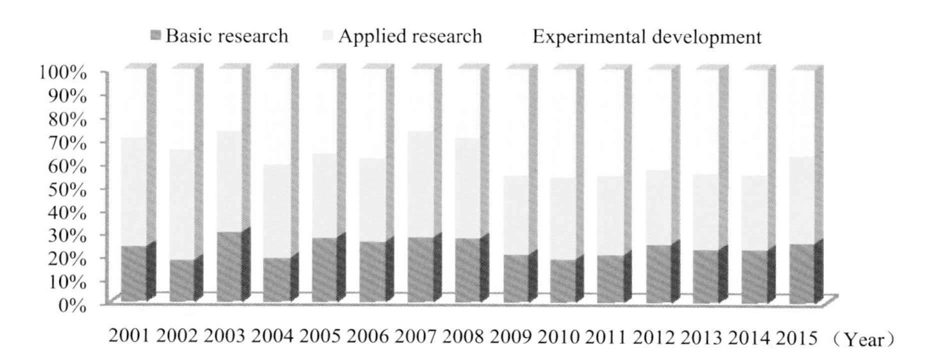

Figure 2-15 R&D Regular Expenses (by Activity Categories) from 2001 to 2015

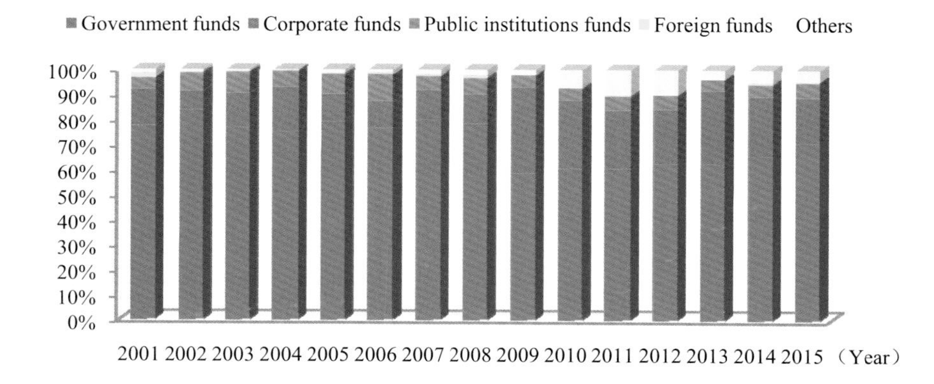

Figure 2-16 Internal Spending of R&D Funds (by Sources) from 2001 to 2015

4. Continuous Growth of Marine Innovation Output Achievements

Knowledge innovation is a core element of national competitiveness. Innovation output means intermediate achievements of various types produced by scientific research and technological innovation activities. The quality and quantity of academic papers and writings reflect the innovation capability of marine science and technology, while patent applications and grants more directly reflect the degree of marine innovation activities and level of technological innovation. Relatively higher ability in marine knowledge diffusion and application is one of the common features of an innovative maritime power.

4.1 The total number of marine S&T papers maintains growth trend.

From 2001 to 2015, the total number of S&T papers in marine field has maintained an overall growth (See Figure 2-17). The number of papers published in 2015 was 3.84 times that of 2001, with an average annual increase of 10.29%. Specifically, CSCD papers has maintained an upward trend in fluctuation, with incremental decrease of published paper appearing in the years 2005, 2008 and 2012. The number of published SCI papers in marine field grows by leaps and bounds, especially during the period of the "*12th Five-Year*" since our country put forward the strategy of building "maritime power", and the number of paper published in international journals has presented an obvious increase trend.

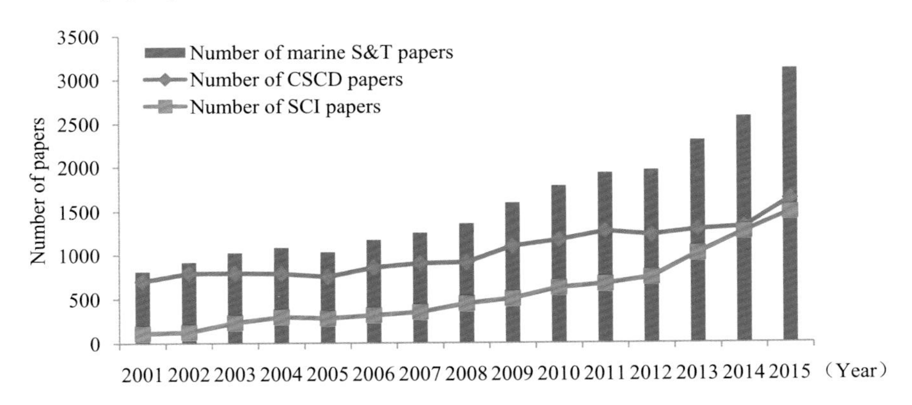

Figure 2-17 Year-over-year Changes in Number of Published Marine S&T Papers in China from 2001 to 2015

In terms of the growth rate of S&T papers every year, the CSCD papers in marine field have presented an upward trend with the exception of 2004, 2005 and 2013. The year 2015 witnessed the maximum growth rate. Except 2005, the number of SCI papers in marine field published every year has shown an increase trend, the maximum growth rate in the number of published academic papers also appeared in 2015 (See Table 2-1).

Table 2-1 Number of Published Marine S&T Papers in China and Analysis of Annual Growth Rate from 2001 to 2015

Year	number of CSCD papers	number of SCI papers	the number of S&T papers	Annual Growth Rate	
				Chinese	English
2001	701	112	813		
2002	791	126	917	13%	13%
2003	792	233	1025	0%	12%
2004	787	298	1085	-1%	6%
2005	751	282	1033	-5%	-5%
2006	856	316	1172	14%	13%
2007	902	351	1253	5%	7%
2008	913	447	1360	1%	9%
2009	1096	503	1599	20%	18%
2010	1167	625	1792	6%	12%
2011	1268	669	1937	9%	8%
2012	1225	744	1969	-3%	2%
2013	1287	1014	2301	5%	17%
2014	1316	1261	2577	2%	12%
2015	1640	1485	3125	25%	21%

4.2 The number of published SCI papers in oceanography sees an obvious increase trend.

The increase presented rapid growth trend especially after 2012 (See Figure 2-18). The number of published papers in 2015 was 1485, 13.26 times that of 2001, with annual growth rate of 21.71%. The proportion in international oceanography papers presents continuous upward trend increasing from 2.16% in 2001 to 20.75% in 2015. Meanwhile, the proportion of the papers with our country as the first country also shows an increase trend, among which the number of SCI papers with our country, as the first country, has presented an linear growth trend between 2011 and 2015 (See Figure 2-20). The year 2015 witness an increase of 90.57%. The scientific papers published abroad are increasing rapidly, clearly indicating that the papers in oceanography of our country are gaining higher positions and obtaining a wide range of international recognition. The rapid growth of published SCI papers in oceanography is far higher than that of SCI papers in international oceanography, indicating on one hand that the starting point of our country is comparatively lower and on the other hand that the innovation and development in marine field are continuously enhanced in recent years.

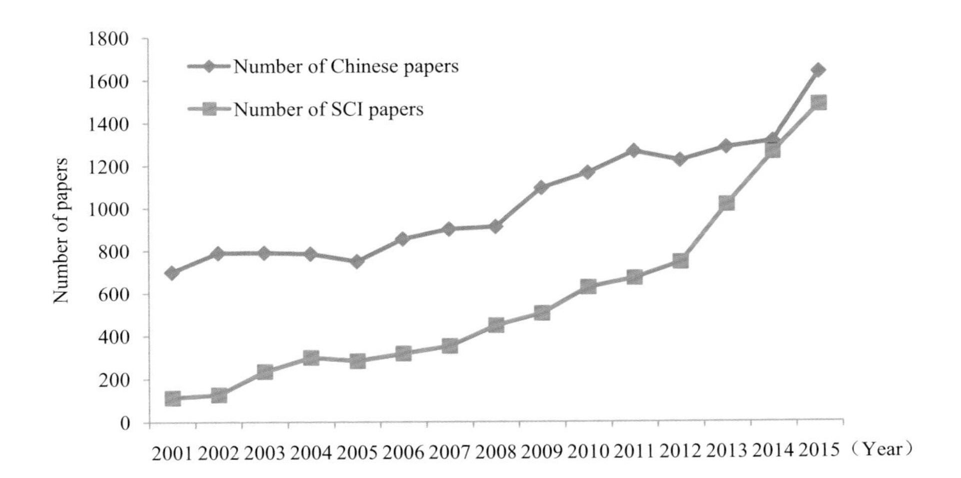

Figure 2-18 Number of SCI Papers in Oceanography of China and Number of Chinese Papers in Oceanography Included in CSCD from 2001 to 2015

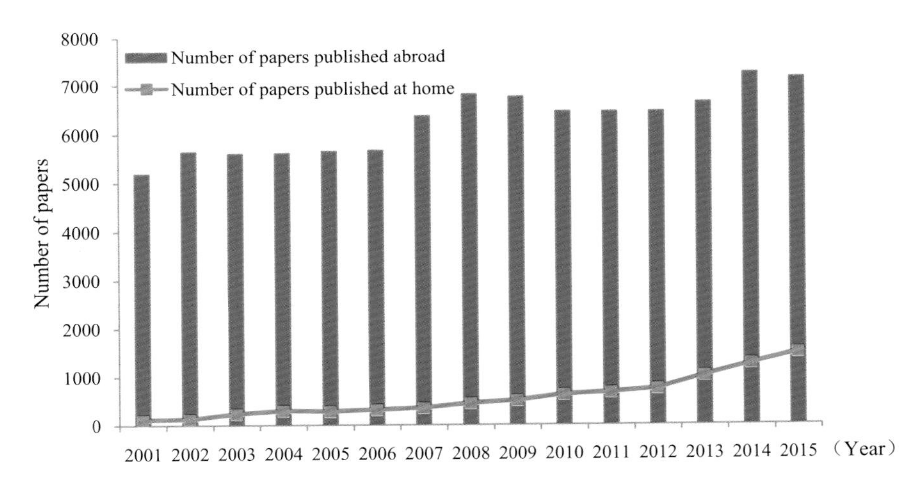

Figure 2-19 Number of SCI Papers in Oceanography Published at Home and Abroad from 2001 to 2015

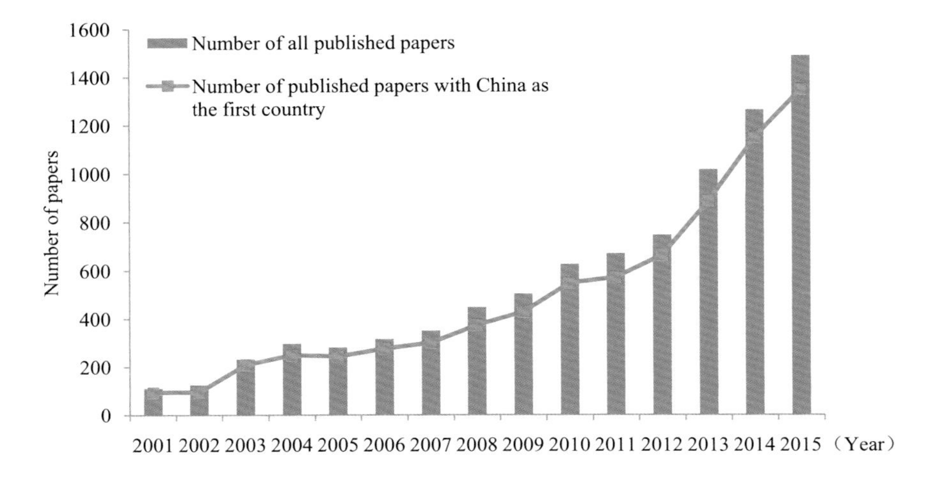

Figure 2-20 Number of All Published SCI Papers in Oceanography of China and Number of Published Papers with China as the First Country from 2001 to 2015

In terms of the citations of SCI papers in oceanography, the total citations of SCI papers in oceanography of our country from 2001 to 2015 are 62869, among which other citations are 61870, with the year 2002 witnessing the highest record of the number of citations per paper which was 26.54 and other citations being 26.47. The citations of papers published at the current years firstly increased and then decreased, with the highest record appearing in 2010. In addition, the H-index of all SCI papers from 2002 to 2010 had been above 30, among which the highest record appeared in 2004 which was 39 (See Table 2-2, Figure 2-21). The average citations of SCI papers per year have presented an upward trend, indicating that oceanography in China is still a hot research discipline (See Figure 2-22).

Table 2-2 H-index of SCI Papers and Citations of Papers Published at the Current Year in Oceanography of China from 2001 to 2015

Year	Total number of citations	H-index	Other citations excluding self-citations	Number of published papers	Number of citations per paper	Number of other citations per paper
2001	1834	24	1829	112	16.38	16.33
2002	3344	30	3335	126	26.54	26.47
2003	3765	31	3750	233	16.16	16.09
2004	5186	39	5152	298	17.40	17.29
2005	3786	30	3770	282	13.43	13.37
2006	4763	37	4727	316	15.07	14.96
2007	4756	36	4736	351	13.55	13.49

Year	Total number of citations	H-index	Other citations excluding self-citations	Number of published papers	Number of citations per paper	Number of other citations per paper
2008	5267	34	5245	447	11.78	11.73
2009	4971	35	4946	503	9.88	9.83
2010	5951	32	5924	625	9.52	9.48
2011	5079	27	5010	669	7.59	7.49
2012	4328	24	4278	744	5.82	5.75
2013	4410	21	4304	1014	4.35	4.24
2014	3357	16	3279	1261	2.66	2.60
2015	2072	10	1585	1485	1.40	1.07

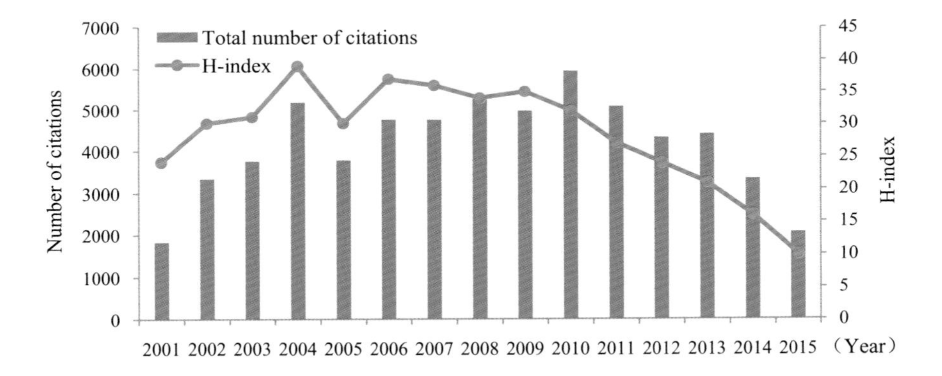

Figure 2-21 Total Number of Citations and H-index of SCI Papers Published Per Year in Oceanography of China

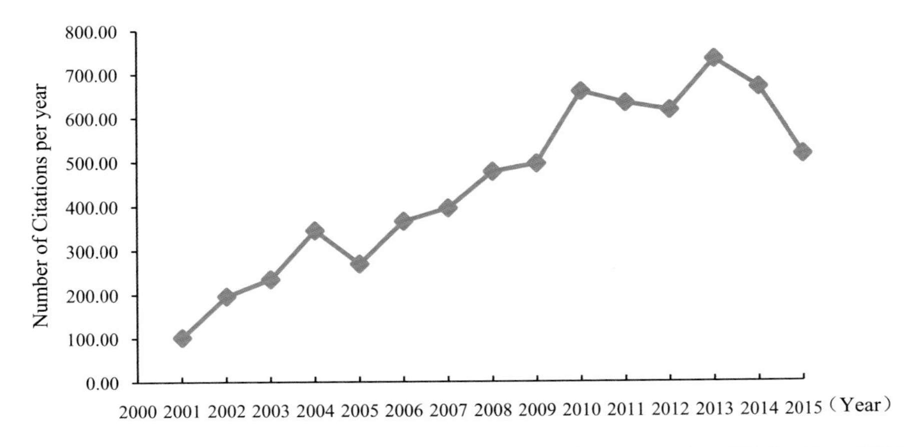

Figure 2-22 Number of Citations of SCI Papers Per Year in Oceanography of China from 2001 to 2015

The papers in oceanography of our country are disproportionally distributed in journals. The number of papers published in three major Chinese SCI journals *ACTA OCEANOLOGICA SINICA, CHINESE JOUNAL OF OCEANLOGY AND LIMNOLOGY* and *JOURNAL OF OCEAN UNIVERSITY OF CHINA* accounts for 49% of all the papers. The number of papers published in journals with IF>2 is 1628, taking up 250% of all papers (See Table2-3). Viewing from the sections of the journals in which the number of SCI papers published is equal to or over 100, the proportion of the papers published in the first section is 27%, and 14% is in the second section, indicating that both the proportion of high-level papers in marine field of our country and scientific research level are relatively high.

Table 2-3 Number of Papers of China Published in Top 20 Journals, Impact Factor & Section from 2001 to 2015

Journal	Impact factor	Section	Number of published papers
ACTA OCEANOLOGICA SINICA	0.631	Q4	1309
CHINESE JOURNAL OF OCEANOLOGY AND LIMNOLOGY	0.547	Q4	1059
CHINA OCEAN ENGINEERING	0.435	Q4	809
OCEAN ENGINEERING	1.488	Q1	535
JOURNAL OF GEOPHYSICAL RESEARCHOCEANS	3.318	Q1	445
JOURNAL OF OCEAN UNIVERSITY OF CHINA	0.509	Q4	424
ESTUARINE COASTAL AND SHELF SCIENCE	2.335	Q1	284
CONTINENTAL SHELF RESEARCH	2.011	Q2	276
MARINE ECOLOGY PROGRESS SERIES	2.011	Q2	157
JOURNAL OF NAVIGATION	1.267	Q1	147
TERRESTRIAL ATMOSPHERIC AND OCEANIC SCIENCES	0.556	Q4	130
APPLIED OCEAN RESEARCH	1.382	Q2	126
JOURNAL OF OCEANOGRAPHY	1.27	Q3	120
MARINE GEOLOGY	2.503	Q1	119
MARINE GEORESOURCES & GEOTECHNOLOGY	0.761	Q2	118
JOURNAL OF ATMOSPHERIC AND OCEANIC TECHNOLOGY	2.159	Q1	117
DEEP-SEA RESEARCH PART II-TOPICAL STUDIES IN OCEANOGRAPHY	2.137	Q2	116
MARINE CHEMISTRY	3.412	Q2	110
JOURNAL OF MARINE SYSTEMS	2.174	Q1	104
ACTA OCEANOLOGICA SINICA	0.631	Q4	1309

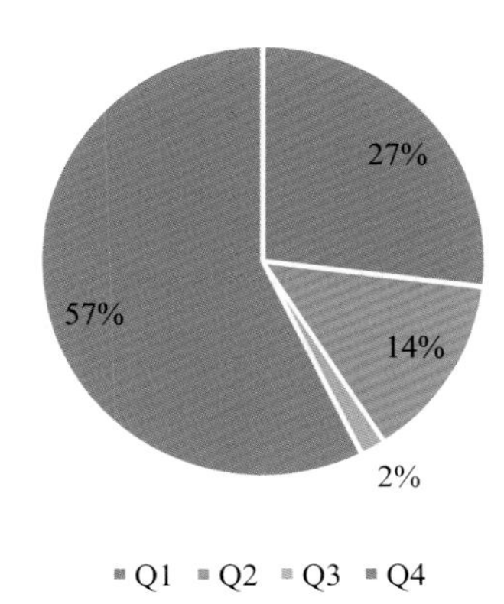

Figure 2-23 Sections of Journals in Which the Number of SCI papers Published≥100

Based on the data source of Chinese Science Citation Database, there are 15492 research papers in oceanography from 2001 to 2015. The journals which published more than 1,000 papers during the period from 2001 to 2015 are *Journal of Oceanography* and *Marine Geology* and *Quaternary Geology*. There are 23 disciplines in which the numbers of papers published in Chinese are more than 30. The disciplines in which the number of papers published is more than 100 include geology, atmospheric sciences, automation technology, basic theory of environmental science, geophysics, and environmental pollution and its prevention and control.

Table 2-4 Major Journals Which Publish CSCD Papers of Marine Science and Technology and Disciplines from 2001 to 2015

Name of journal	Numbers of published papers	Discipline	Numbers of published papers
Acta Oceanologica Sinica	1088	Geology	422
Marine Geology & Quaternary Geology	1068	Atmospheric sciences (Meteorology)	247
Chinese Journal of Oceanology and Limnology	865	Automation, computer technology	182
Marine Sciences	842	Basic theory of environmental sciences	148
Marine Sciences bulletin	777	Geophysics	104
Journal of Tropical Oceanography	767	Environmental pollution and control	102
Acta Oceanologica Sinica	684	General Biology	94
The Ocean Engineering	644	Mechanics	92
Periodical of Ocean University of China	543	Hydraulic engineering	89
Marine Environmental Science	524	Physical geography	80

Name of journal	Numbers of published papers	Discipline	Numbers of published papers
Advances in Marine Science	497	Topography	71
Oceanologia et Limnologia Sinica	426	Chemical industry	65
Journal of Oceanography In Taiwan Strait	399	Fishery sciences	62
Advances in Earth Science	314	Chemistry	58
Transactions of Oceanology and Limnology	288	Maritime transport	56
Technology of Water Treatment	238	Social sciences	52
Journal of Applied Oceanography	207	Environmental quality assessment and Environmental monitoring	51
Chinese Science Bulletin	192	Oil and gas industry	47
Journal of Lake Science	189	Physics	44
Quaternary Sciences	138	Mathematics	42
Progress in Geophysics	137	Architecture	35
Chinese Journal of Geophysics	137	Energy and power engineering	30

With regard to cooperations with other countries and Taiwan, there are 38 countries or regions with more than 5 coauthored SCI papers in Oceanography, with USA ranking the first among the top 5 countries in number of cooperations, followed by Britain, Japan, Australia and Canada (See Figure 2-24).

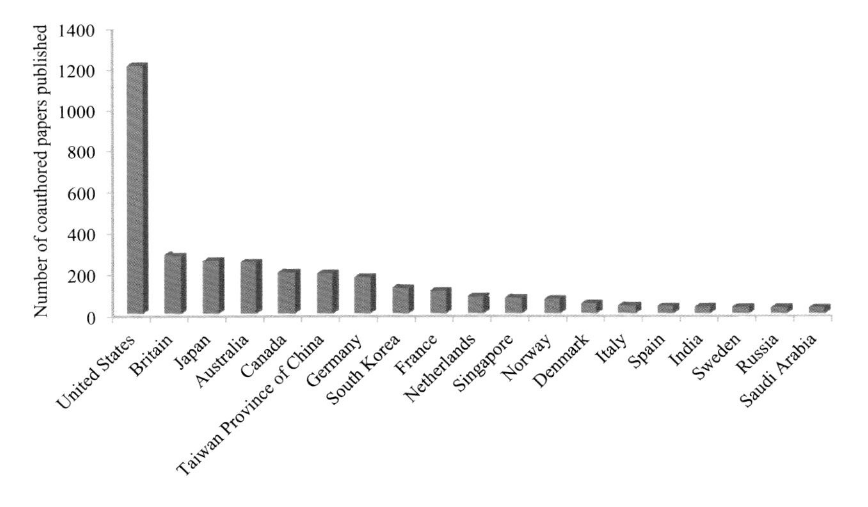

Figure 2-24 Cooperation Relationship of SCI Papers in Oceanography

4.3 The categories of marine S&T publications increase significantly.

From 2001 to 2015, the categories of marine S&T publications by China's marine scientific research institutes have shown a significant growth trend (See Figure 2-25), with an average annual growth rate of 19.10% and the year 2013 witnessing the highest record. Among them, the categories of marine S&T publications had maintained a stable growth from 2001 to 2005, with an average growth rate of 7.93%; the 2006-2007 and 2008-2009 periods had witnessed a rapid growth in the categories of marine S&T publications, with growth rates of 102.86% and 78.57% respectively; after 2010, the annual growth rate of the categories of marine S&T publications has maintained at 9.06%.

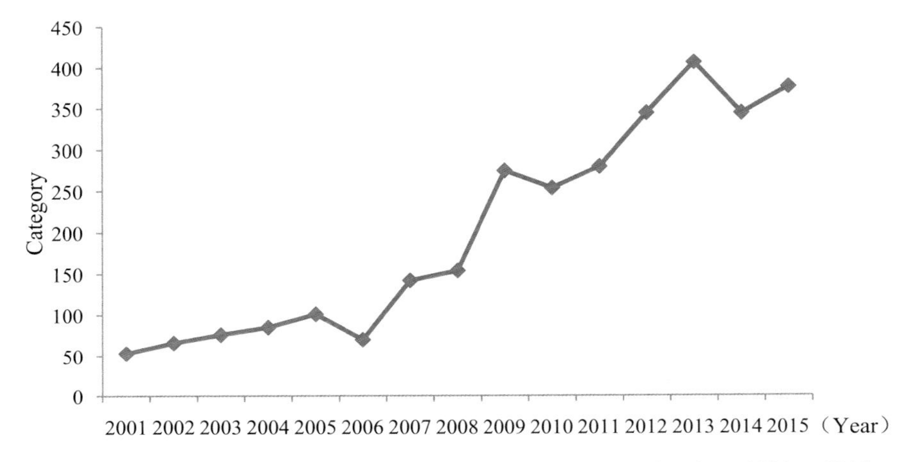

Figure 2-25 Categories of Marine S&T Publications of China from 2001 to 2015

4.4 The number of patent applications in marine field witnesses strong growth.

The number of patent applications in marine field has been on the rise year by year, increasing by 17 times from 248 in 2001 to 4333 in 2015. Before 2006, the annual increase in terms of the number of patents had maintained a stable trend with little fluctuation. After 2007, the increase has maintained at above 20%, ushering in a period of rapid development (See Figure 2-26).

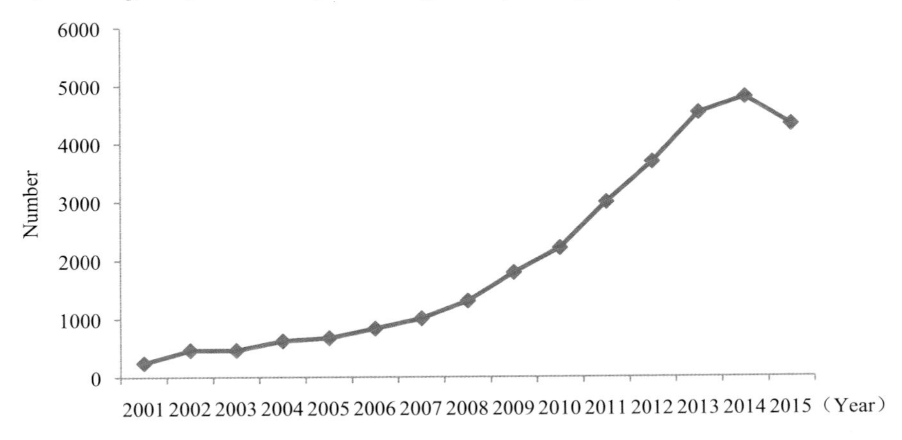

Figure 2-26 Number of Patent Applications in Marine Field of China from 2001 to 2015

On the basis of increasing number of patent applications, the number of valid patents (including pending patents and granted patents) accounts for 64.5% and the number of unpaid patents accounts for 20.8% (See Figure 2-27).

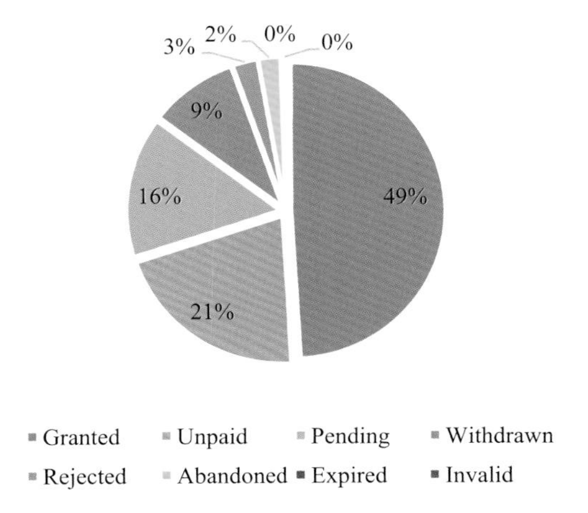

Figure 2-27 Legal Status of Patents in the Marine Field of China from 2001 to 2015

Invention patents account for 60% of patents in the marine field of our country (See Figure 2-28), reflecting to a certain degree the greater potential of technological innovation in the marine field of our country. The proportions of utility model and design patents are very low, indicating in a certain way the small number of mature products in the marine field of our country.

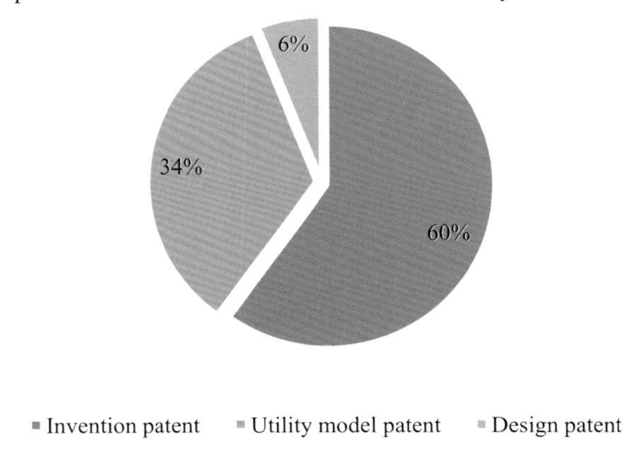

Figure 2-28 Proportion of Patent Categories in the Marine Field of China from 2001 to 2015

There is little change in the growth rate of design patents while the invention and utility model patents show a rapid increase.

Among the top 15 major application institutes which have obtained patents in the marine field of China, there are 3 enterprises (all are from different subsidiaries of China National Offshore Oil

Corporation), 8 universities and 3 research institutes (See Figure 2-30).

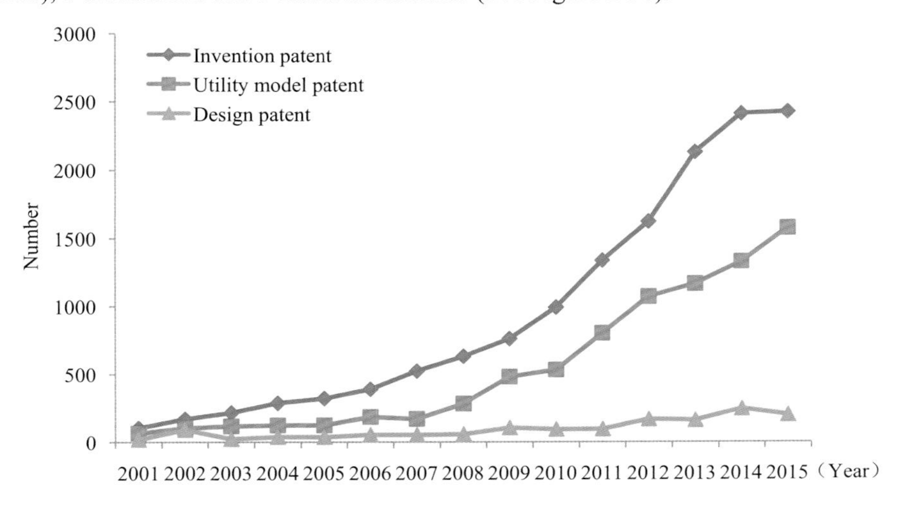

Figure 2-29 Trend of Patent Categories in the Marine Field of China from 2001 to 2015

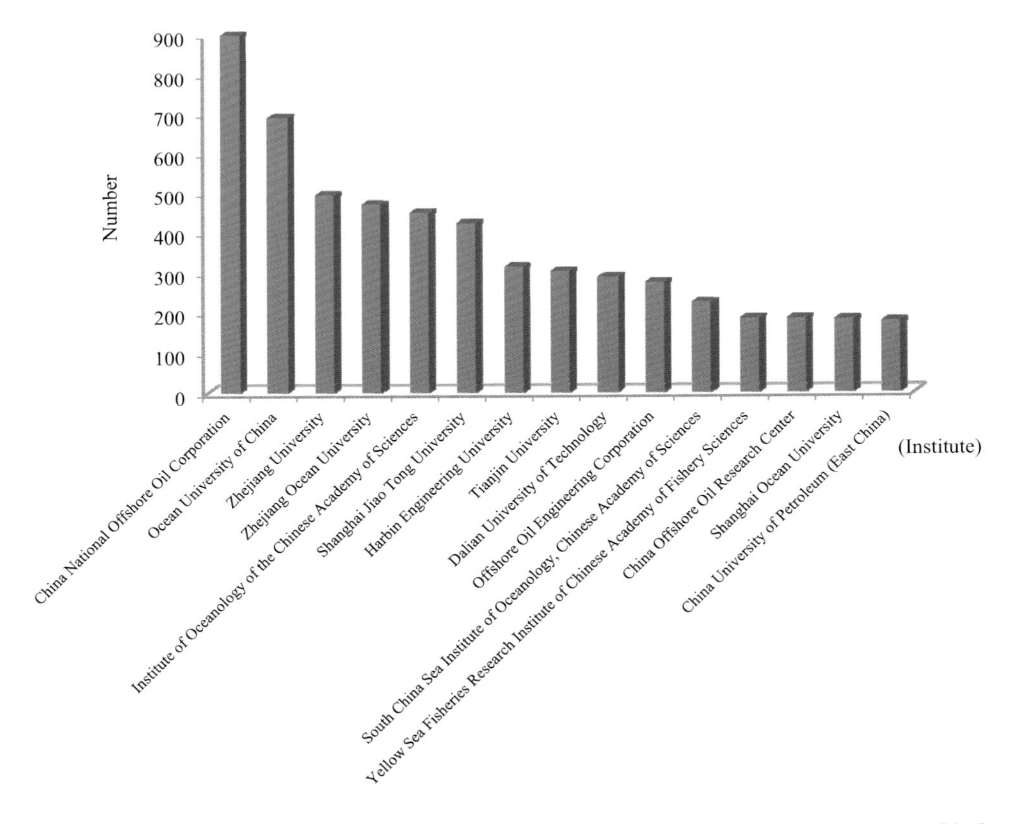

Figure 2-30 Major Patent Application Institutes in the Marine Field of China from 2001 to 2015

Among the top 15 patent applicants in terms of application numbers, most of their patent activity periods are more than 10 years and average patent age is above 5 years (see Table 2-5). The average number of people who invented every patent is 1.58 and China National Offshore Oil Corporation reaches 2.48.

Table 2-5 Comprehensive Index of Applicants

Applicants	Number of patents	Percentage	Comparison of the applicants' R&D capability		
			Patent activity period	Number of inventors	Average patent age
China National Offshore Oil Corporation	936	3.12	13	2323	5.1
Ocean University of China	690	2.30	14	892	6.8
Zhejiang University	494	1.65	14	600	6.6
Zhejiang Ocean University	470	1.57	10	412	4.3
Institute of Oceanology of the Chinese Academy of Sciences	450	1.50	15	417	6.8
Shanghai Jiao Tong University	424	1.41	15	408	7.3
Harbin Engineering University	315	1.05	14	620	4.9
Tianjin University	305	1.02	15	411	5.3
Dalian University of Technology	289	0.96	14	402	5.0
Offshore Oil Engineering Corporation	276	0.92	11	949	4.9
South China Sea Institute of Oceanology,Chinese Academy of Sciences	224	0.75	14	229	6.2
Yellow Sea Fisheries Research Institute, Chinese Academy Of Fishery Sciences	185	0.62	14	222	5.9
China Offshore Oil Research Center	185	0.62	8	305	7.4
Shanghai Ocean University	182	0.61	7	378	4.9
China University of Petroleum	177	0.59	11	347	3.6

Among the top 15 patent applicants in terms of patent categories, most are invention patents and only one is design patent belonging to Harbin Engineering University (See Figure 2-31).

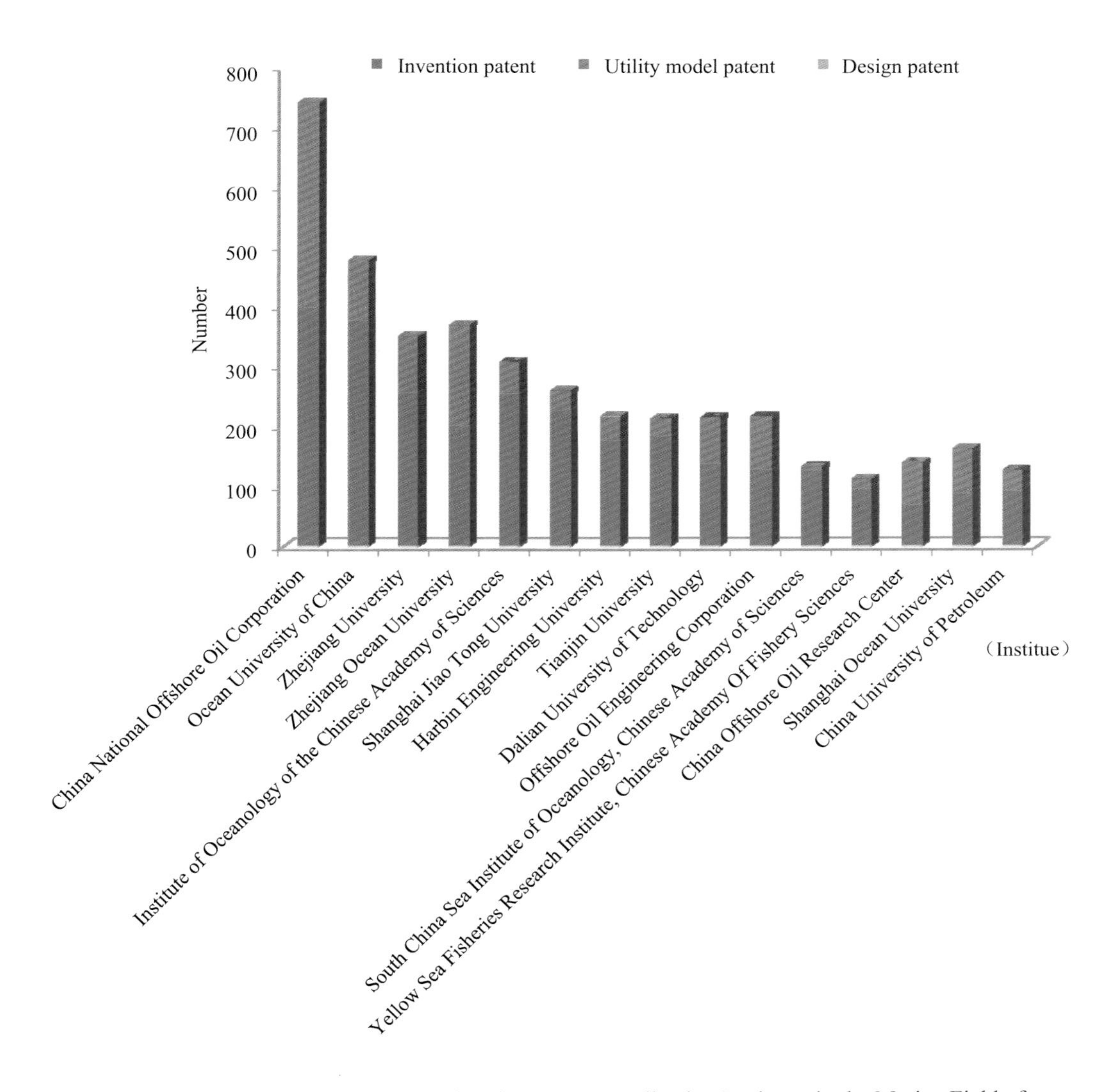

Figure 2-31 Patent Categories of Major Patents Application Institutes in the Marine Field of China from 2001 to 2015

Among the major patents application provinces in the marine field of China, Shandong ranks the first due to its comparatively larger number of marine-related scientific research institutes and universities, Beijing ranks the fourth and the number of applications of Fujian province is relatively small (See Figure 2-32).

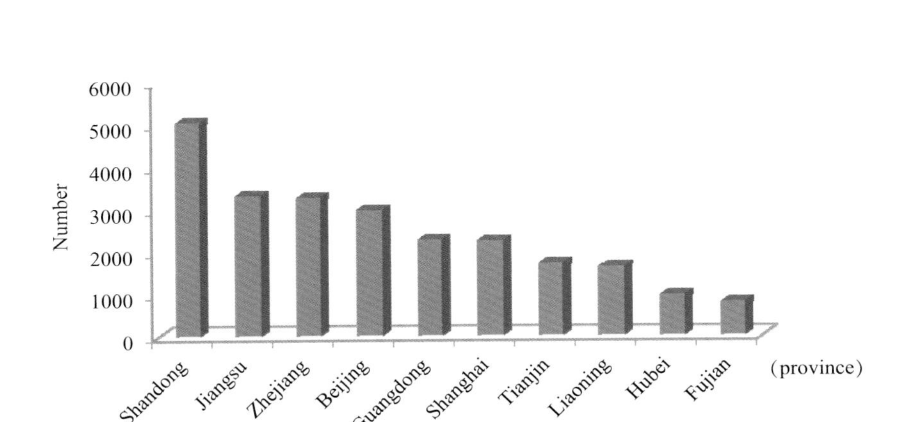

Figure 2-32 Major Patents Application Provinces in the Marine Field of China from 2001 to 2015

Judging from the patents application categories of major provinces, Zhejiang, Jiangsu and Guangdong boost most of the design patents, while the proportion of invention patents and utility model patents of various provinces is similar (See Figure 2-33).

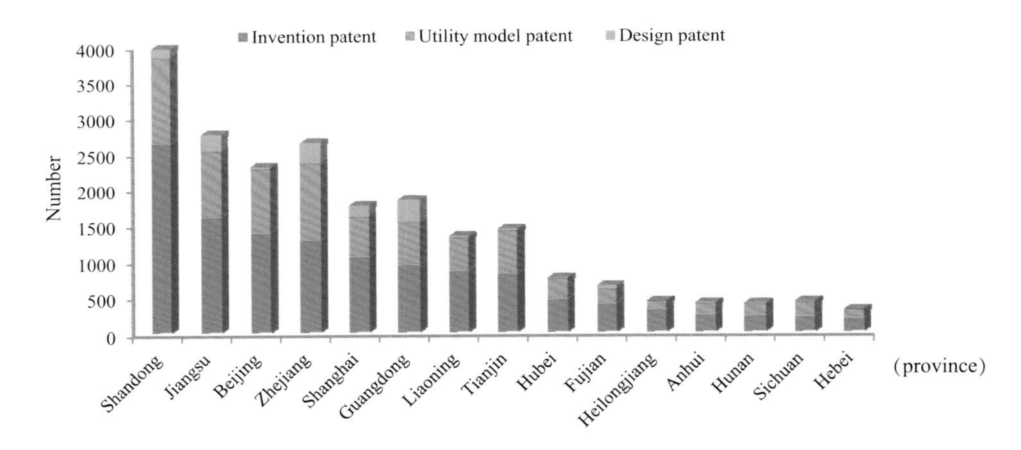

Figure 2-33 Patent Categories of Major Patents Application Provinces in the Marine Field of China from 2001 to 2015

The top 15 categories of patents with relatively higher occurrence frequency in the marine field of China are as follows: C02F (sewage and sludge contamination treatment), G01N (testing or analyzing materials by means of chemical or physical property of the materials), A01K (fisheries management; fish farming), B63B (ship or other waterborne vessels; marine equipment), F03B (hydraulic machinery or hydraulic motor), E21B (soil or rock drilling), A61K (medical preparations and formulations), C09D (coating composition, such as colored paint, varnish or natural lacquer; filling slurry; chemical coating or printing ink remover; printing ink; correction fluid; wood stain; slurry or solids used for staining or printing; application of raw materials for this purpose), A61P (curative property of chemical compounds or pharmaceutical preparation), C12N (micro-organisms or enzymes), E02B(preparation for grinding grain; refining grain into goods by hulling), C12R (index related to the subdivision of microorganisms from C12C to C12Q or C12S), B01D (separation), F16L (pipe; pipe joint or pipe fitting; pipe, cable or protective pipe support; general adiabatic method), E02D (Foundation; excavation; embankment; underground or underwater structure) (See Figure 2-34 and 2-35).

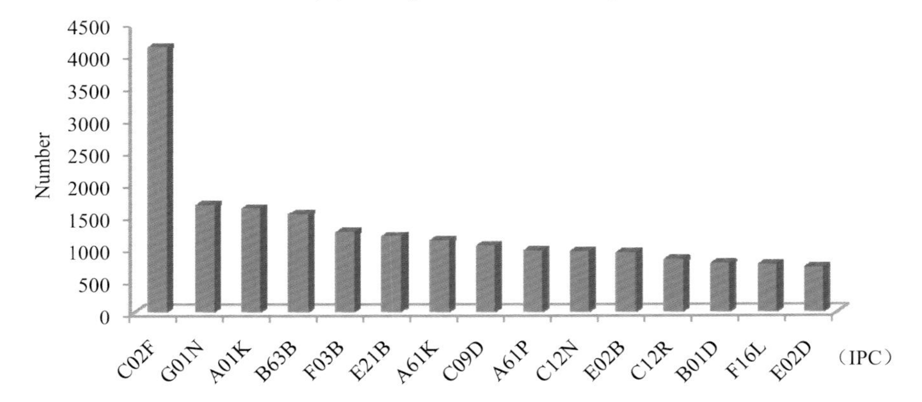

Figure 2-34 Major Classification Numbers of Patents in the Marine Field of China from 2001 to 2015

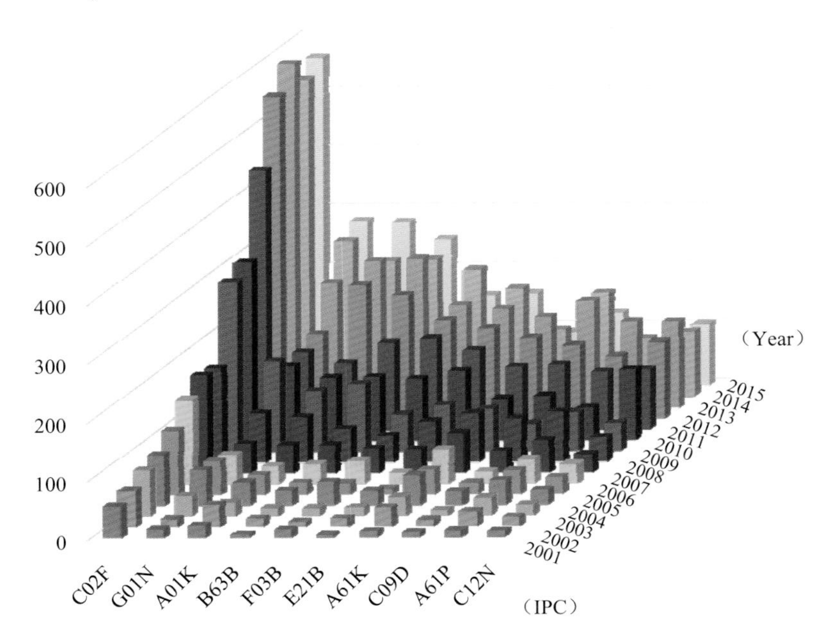

Figure 2-35 Number of the Top 10 Patents Application by IPC in the Marine Field of China from 2001 to 2015

4.5 China dominates obvious advantageous position in international marine patents.

According to the data of marine patents from 2001 to 2015 retrieved by DII database, the application number of patents in China ranks first, surpassing other countries by a far distance, equivalent to the total of South Korea, Japan and the United States (See Figure 2-36).

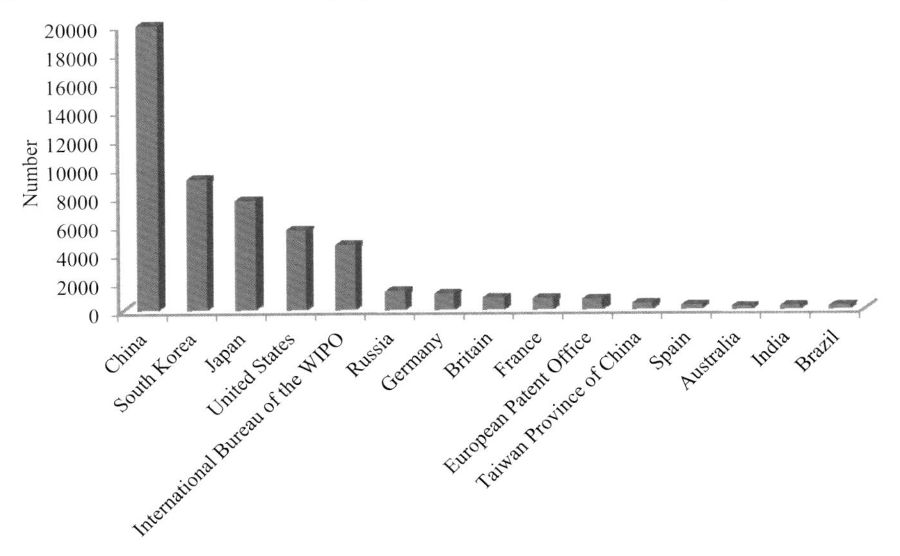

Figure 2-36 Number of Patents Application of Various Countries in Marine Field from 2001-2015

In terms of the number of patents, the growth trend of China has been basically consistent with that of the world and the proportion in world patents has maintained stable increase (See Figure 2-37), increasing from 5.40% in 2001 to 75.90% in 2015 (including cooperative patents with many countries).

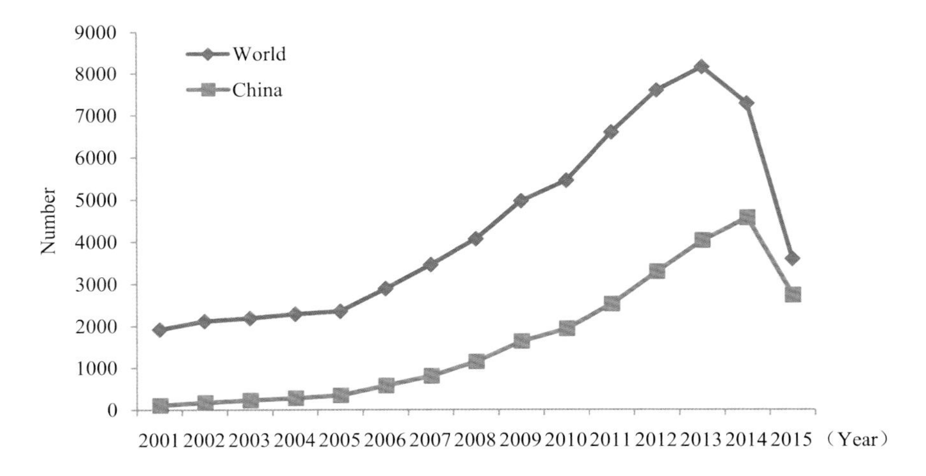

Figure 2-37 Number of Marine Patents Application in the World from 2001 to 2015

The world's top 15 marine patents by IPC are: B63B (ship and other waterborne vessels; marine equipment), C02F (sewage and sludge contamination treatment), A01K (animal husbandry; the management of birds, fish, and insects; fishing; raising or breeding animals which are not included by other breeds; new breeds of animals), A23L(food, foodstuff or non-alcoholic drinks which are not included in A21D or from A23B to A23J), E02B (water conservancy project), F03B (hydraulic machinery or hydraulic motor), A61K (medical preparations and formulations), B01D (separation), A61P (curative property of chemical compounds or pharmaceutical preparation), B63H (the propulsion device or steering device for ships), E21B (soil or rock drilling), G01N (testing or analyzing materials by means of chemical or physical property of the materials), G01V(geophysics; gravity measurement; detection of matter or object; tracer), E02D (foundation; excavation; embankment), G01S (radio orientation; radio navigation; using radio wave for measuring distance or speed; measurement or speed; using radio wave reflection or the positioning or presence of re-radiation for detection; using similar devices of other waves), E02D (Foundation; excavation; embankment; underground or underwater structure) (See Figure 2-38).

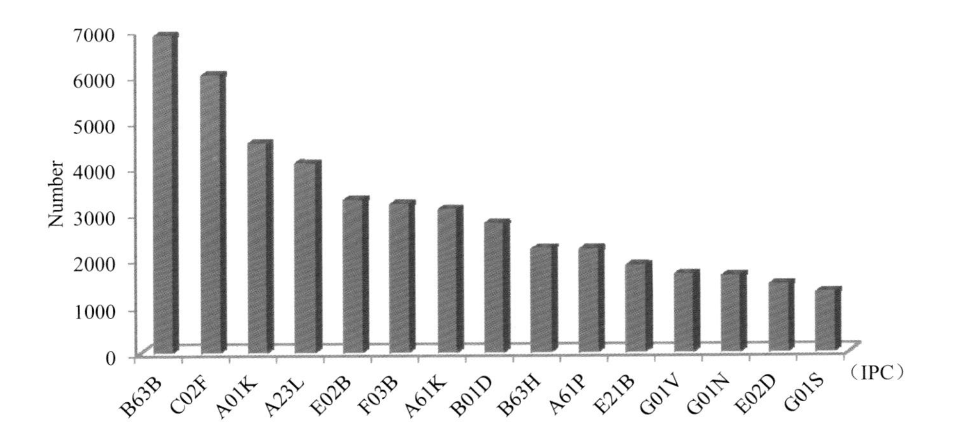

Figure 2-38 IPC Number of the World Marine Patents Application from 2001 to 2015

Among the world's top 15 institutes in terms of the number of marine patents application, there are 4 in China (See Figure 2-39).

Overall, patents in the marine field of China are developing fast and have certain quantitative advantages in the world marine patents. The R&D strengths of patents in the marine field of China are evenly distributed at enterprises, universities and research institutes. But the major applicants are comparatively concentrated and the R&D strengths of enterprises are especially focused on China National Offshore Oil Corporation. The patents in the marine field of China are mainly distributed in such fields as fisheries, pharmaceutical industries and mineral extraction but most are at the stage of original technology exploration (such as more invention patents) and less found in new products, food and high-tech (such as few design patents). The future development of marine patents should not only depend on the exploration of new technology and new direction, but more importantly be focused on the transfer and transformation of existing patent industrialization.

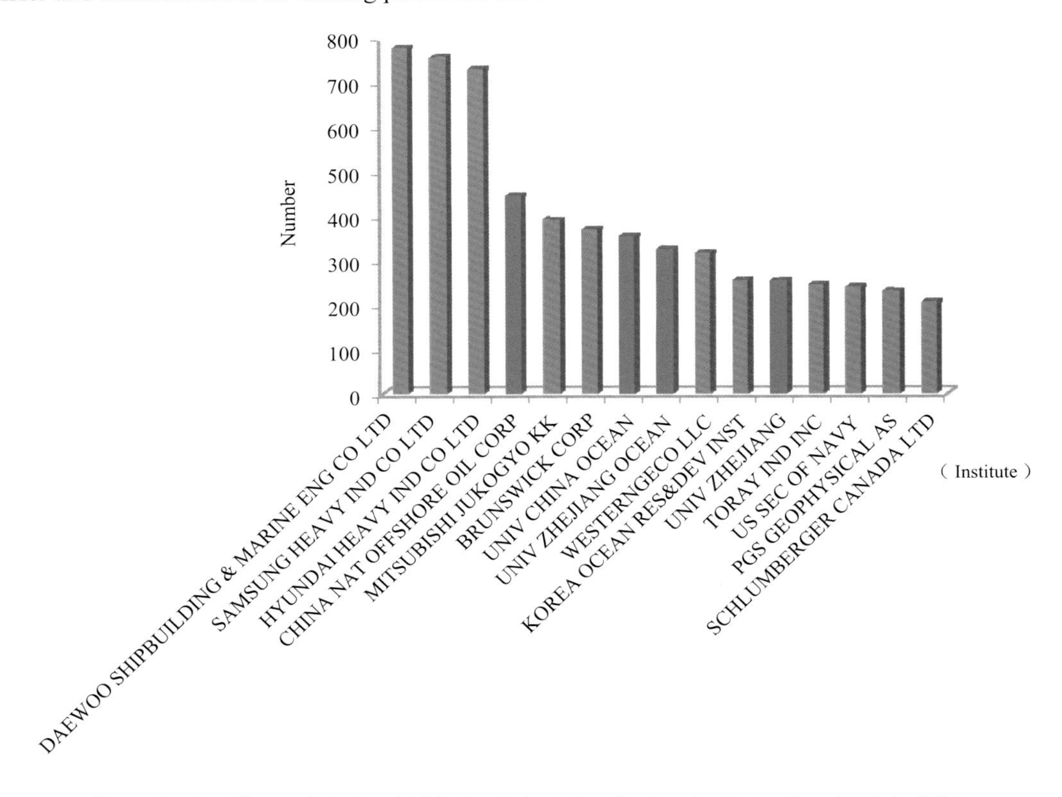

Figure 2-39 The world's Top 15 Marine Patents Application Institutes from 2001 to 2015

5. Sound Conditions for Marine Innovation Development in Higher Institutions

Higher institutions play a decisive role in national innovation development. In recent years, both the input of marine innovation resources and the output of marine innovation achievements at higher institutions of China have gradually been boosted and marine innovation has developed healthily. It should be noted that the data used in this part are extracted on the basis of marine-related higher institutions and marine-related disciplines and then obtained by weighted summation according to their marine-related coefficient of proportionality (See Appendix 10 for the list of marine-related higher institutions and marine-related coefficient proportionality and Appendix 11 for marine-related disciplines).

5.1 The structure of human resources for marine innovation in higher institutions has been gradually optimized.

"The teaching and research staff at the higher institutions" refer to the staff on the payroll of higher institutions who are engaged in teaching, R&D, and application of R&D results at the junior college level and above, plus the technology service staff and the personnel who provide services for the above work during the statistical year, including foreign experts and visiting scholars outside the higher education system who have been engaged in scientific research activities for more than one month in the aggregate, within the statistical year. From 2009 to 2015, the number of teaching and research staff at the Chinese higher institutions has presented an overall growth trend with slight decline in 2015, of whom, the number of scientists, engineers and staff with senior professional titles also has shown an upward trend, with slight fluctuations in the number of scientists and engineers engaged in teaching and research (See Figure 2-40) and the proportion of staff with senior professional titles in the teaching and research field has risen from 37.50% to 41.38%.

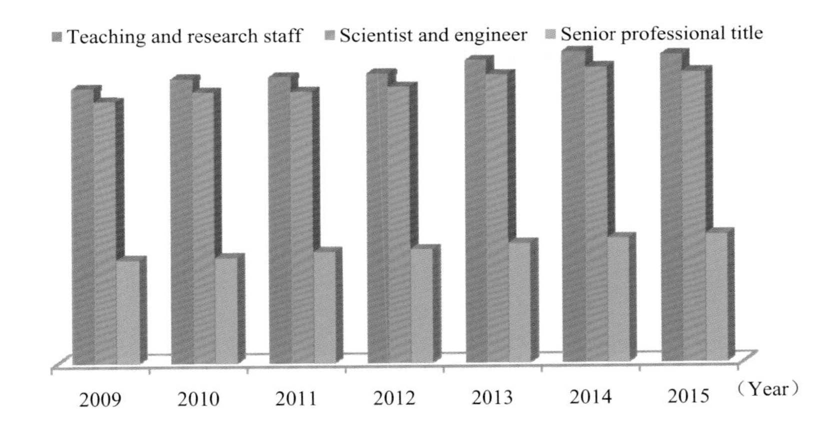

Figure 2-40 Growth Trend of Teaching and Research Staff at Marine-related Higher Institutions of China from 2009 to 2015

"R&D personnel at the higher institutes" refers to the teaching and scientific research personnel whose time spent in R&D accounts for more than 10% of the total time spent in their own teaching and scientific research within the statistical year. From 2009 to 2015, the number of R&D staff at the marine-related colleges and universities in China has gradually increased, of whom, the number of scientists and engineers and the staff with senior professional titles has also shown an upward trend, the proportion of scientists and engineers among the R&D staff has been rising in fluctuation, increasing from 95.99% to 96.459% and there has been slight fluctuations in the proportion of staff with senior professional titles among the R&D staff (See Figure 2-41).

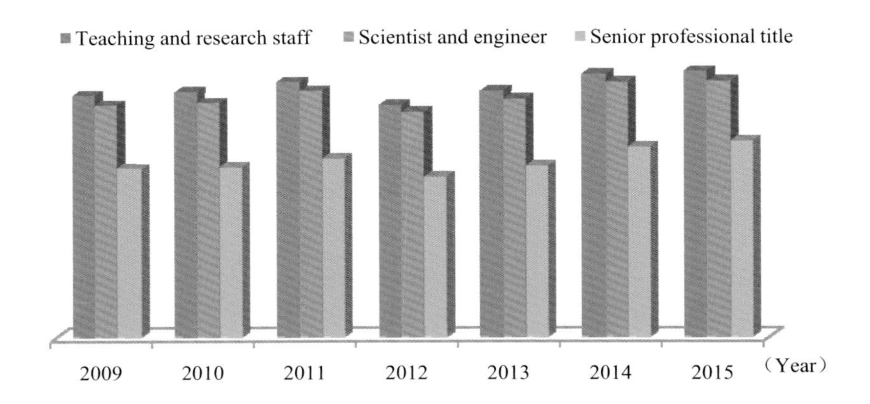

Figure 2-41 Growth Trend of R&D Personnel at Marine-related Higher Institution of China from 2009 to 2015

5.2 The input in marine innovation of higher institutions has been gradually increased.

From 2009 to 2015, China's marine-related colleges and universities have made increasing investment in science and technology funds, with an average annual growth rate of 12.30%. From 2009 to 2015, the government funds have shown a growing trend, with an average annual growth rate of 13.37%. From 2009 to 2015, the internal spending of marine-related higher institutions in China has shown a significant growth (See Figure 2-42), and the internal spending in 2015 was 61.22 times that of 2009.

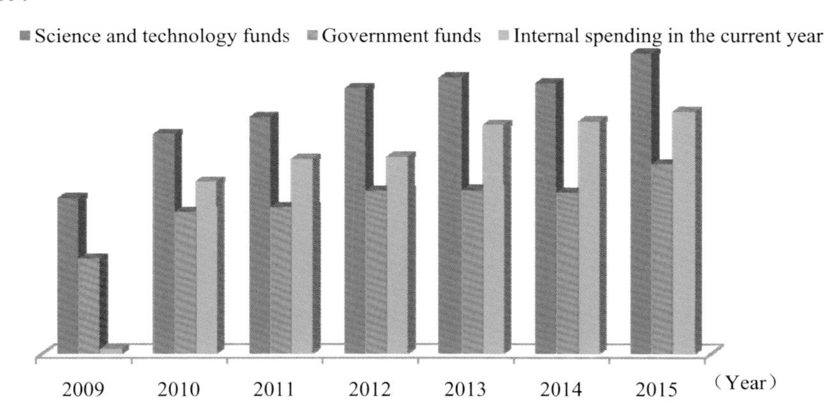

Figure 2-42 Trend of Revenues and Expenditures (thousand yuan) of Science and Technology Funds at Marine-related Higher Institutions of China from 2009 to 2015

From 2009 to 2015, the total number of S&T projects at the marine-related higher institutions of China has shown gradual increase, with an average annual growth rate of 6.26%; the number of staff involved in S&T projects in the current year has shown an overall upward trend, with an average annual growth rate of 1.29% (See Figure 2-43). From 2009 to 2015, the appropriated funds and expenditure for S&T projects at the marine-related higher institutions in China in the current year have increased year by year with the average annual growth rate of appropriated funds in the current year reaching 11.59% and the average annual growth rate of expenditure in the current year reaching 11.92% (See Figure 2-44).

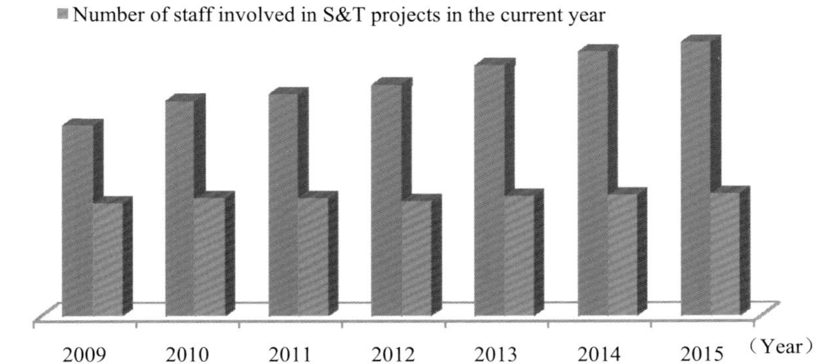

Figure 2-43 Trend of the Total Number of S&T Projects and the Number of Staff Involved S&T Projects in the Current Year at Marine-related Higher Institutions of China from 2009 to 2015

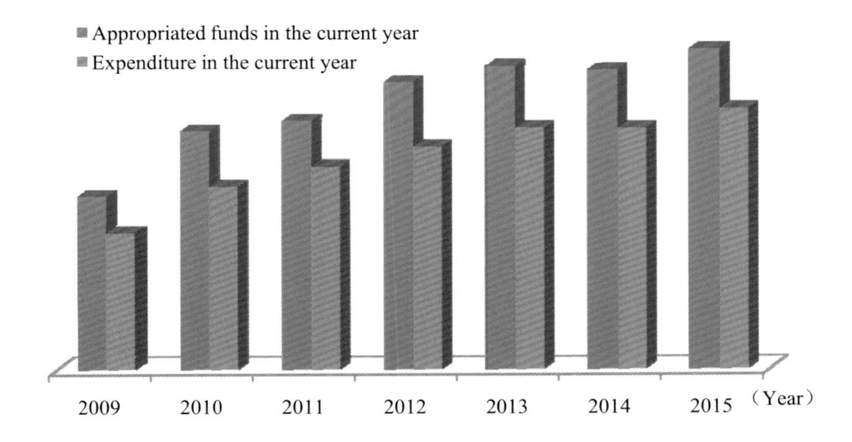

Figure 2-44 Trend of the Appropriated Funds (thousand yuan) and Expenditure (thousand yuan) for S&T Projects at the Marine-related Higher Institutions of China in the Current Year from 2009 to 2015

5.3 The marine innovation output of higher institutions has been gradually increased.

From 2009 to 2015, the number of academic papers published regarding the S&T achievements at China's marine-related higher institutions has gradually increased (See Figure 2-45), with an average annual growth rate of 5.42%, of which the number of academic papers published in foreign academic journals has witnessed obvious growth, with an average annual growth rate of 11.95%. The number of contracts signed for technology transfer has increased year by year (See Figure 2-46), with an average annual growth rate of 18.48% , of which, the period of 2009-2010 had witnessed the most rapid growth with an annual growth rate of 67.60%.

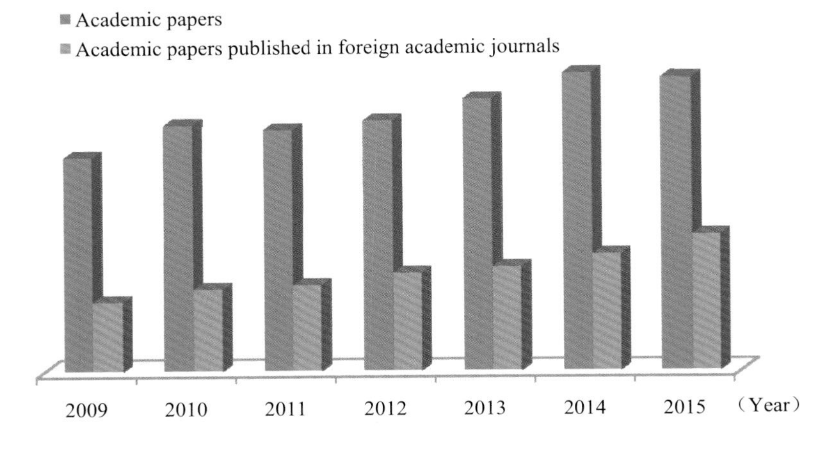

Figure 2-45 Growth Trend of the Number of Academic Papers Published Regarding the S&T Achievements at Marine-related Higher Institutions of China from 2009 to 2015

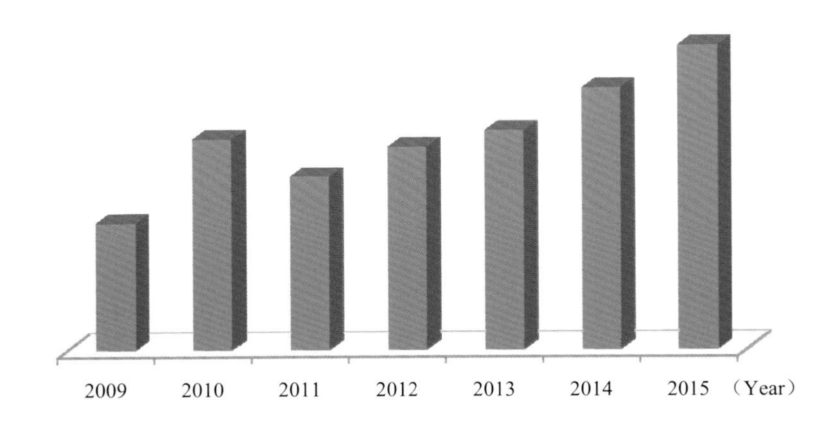

Figure 2-46 Number of Technology Transfer Contracts at Marine-related Higher Institutions of China from 2009 to 2015

5.4 The marine scientific research institutes of higher institutions develops steadily.

From 2012 to 2015, the number of employees at marine-related scientific research institutes of higher institutions in China has increased gradually (See Figure 2-47), of whom, the number of doctoral and master graduates has also presented growth trend, meanwhile, the proportion of doctoral graduates has increased from 51.76% to 56.22%, and the proportion of master graduates has increased from 27.56% to 28.97% (See Figure 2-48).

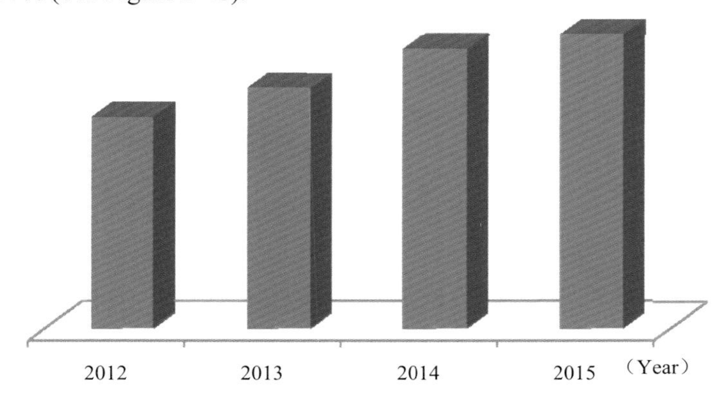

Figure 2-47 Growth Tendency of the Employees at Marine-related Scientific Research Institutes of Higher Institutions in China from 2012 to 2015

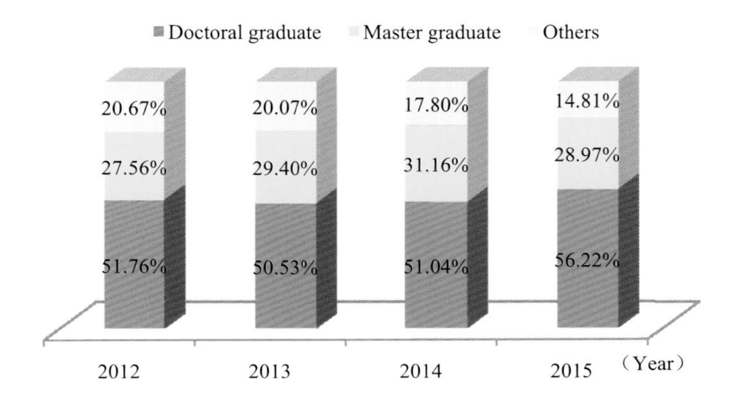

Figure 2-48 Academic Qualification Structure of Employees at Marine-related Scientific Research Institutes of Higher Institutions in China from 2012 to 2015

From 2012 to 2015, the number of S&T personnel at marine-related scientific research institutes of higher institutions in China has gradually increased (See Figure 2-49), of whom, the proportion of senior professional title holders has decreased from 60.00% to 58.85%, the proportion of intermediate professional title holders has increased from 28.46% to 32.06%, and the proportion of junior professional title holders has decreased from 7.76% to 6.89% (See Figure 2-50).

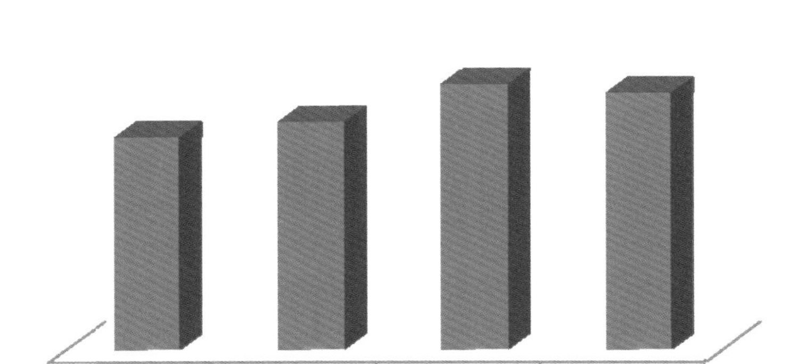

Figure 2-49 Trend of S&T Personnel at Marine-related Scientific Research Institutes of Higher Institutions in China from 2012 to 2015

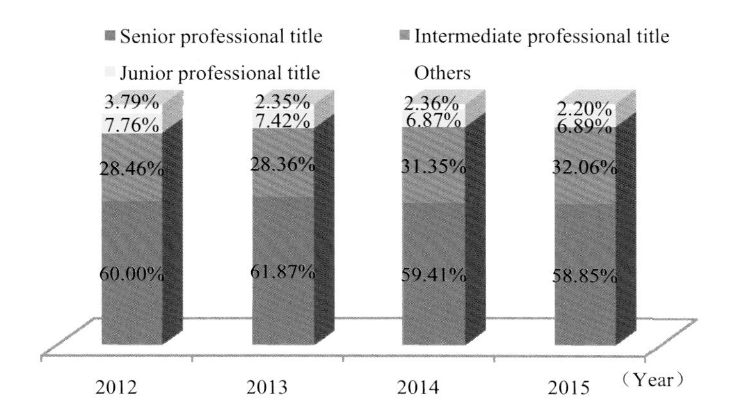

Figure 2-50 Professional Title Structure of S&T Personnel at Marine-related Scientific Research Institutes of Higher institutions in China from 2012 to 2015

From 2012 to 2015, the input of S&T funds at marine-related scientific research institutes of higher institutions in China has gradually increased (See Figure 2-51), the internal spending in 2015 increased by 61.07% compared to 2012, of which the R&D spending in 2015 increased by 69.01% compared to 2012.

From 2012 to 2015, the total number of projects undertaken by marine-related scientific research institutes of higher institutions in China has been on a gradual rise (See Figure 2-52), increasing by 32.47% in 2015 compared to 2012.

From 2012 to 2015, the original value of fixed assets at marine-related scientific research institutes of higher institutions in China has grown year by year (See Figure 2-53), increasing by 37.66% in 2015 compared to 2012. The original value of instruments and equipment increased by 36.90% in 2015 compared to 2012, and the original value of inward instruments and equipment increased by 51.98% compared to 2012.

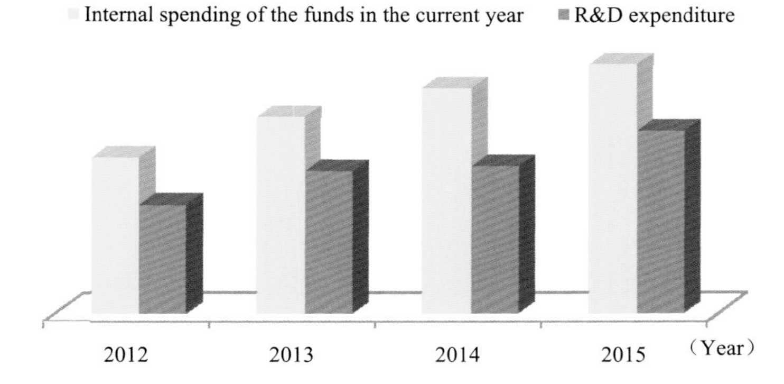

Figure 2-51 Trend of Funds Expenditure (thousand yuan) at Marine-related Scientific Research Institutes of Higher institutions in China from 2012 to 2015

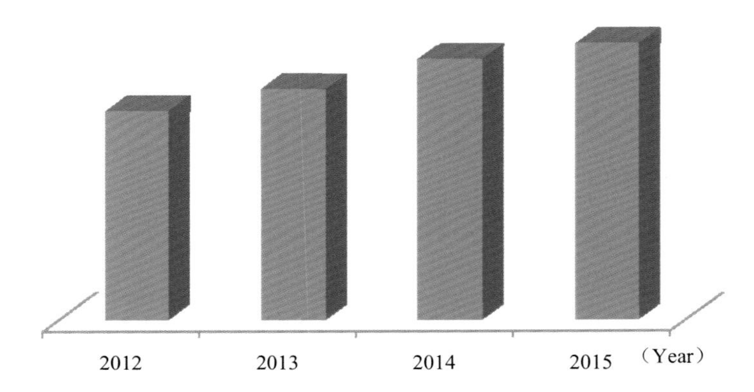

Figure 2-52 Growth Trend of the Number of Projects Undertaken by Marine-related Scientific Research Institutes of Higher institutions in China from 2012 to 2015

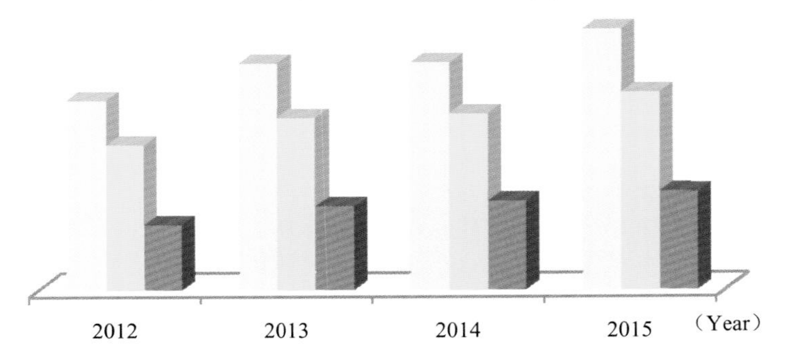

Figure 2-53 Growth Trend of the Original Value of Fixed Assets (thousand yuan) at Marine-related Scientific Research Institutes of Higher institutions in China from 2012 to 2015

6. Steady Increase of Marine S&T Contributions to Economic Development

In recent years, marine innovation work has progressed solidly and a large number of achievements have emerged, thereby advancing the cause of marine development. In this process, the capacity of marine technology to serve marine economic and social development has increased continuously, and the role of technological innovation has become more pronounced in promoting transformation of achievements.

The contribution rate of marine S&T progress witnesses a steady growth. The contribution rate of marine S&T progress refers to the share of the contribution made by marine S&T progress to marine economic growth. It is not only an important indicator for measuring the degree of marine S&T progress, but also a comprehensive indicator for measuring the competitive strength of marine science and technology and the transformation level of marine science and technology into practical productive forces. "*Outline of the National '12th Five-Year Plan' for the Marine Scientific and Technological Development*" explicitly proposed that "the contribution rate of marine science and technology to the marine economy will have reached by more than 60% by 2015". According to the data of "*China Marine Statistical Yearbook*" over the years and improved *Solow Residual Method* on the basis of weighting (See Appendix 5 and 6 for calculation process), the contribution rate of marine S&T progress during the "*11th Five-year Plan*" period (2006-2010) and the early stage of the "*12th Five-year Plan*" period (See Table 2-6) has been measured and calculated.

Table 2-6 Contribution Rate of Marine S&T Progress in China (%)

Year	Growth rate of output	Growth rate of capital	Growth rate of labor	Contribution rate of marine S&T progress (E)
2006—2010	12.86	10.10	4.05	54.4
2006—2015	10.97	6.74	2.72	64.2

As can be seen from Table 2-6, the contribution rate of marine S&T progress during the "*11th Five-year Plan*" period was 54.40%, which reached 64.20% during the 2006-2015 period. In other words, during this period, gross marine product in China has grown at a rate of 13.07%, of which 64.2% come from the contribution made by marine S&T development. The objectives proposed in the "*Outline of the National '12th Five-Year Plan' for the Marine Scientific and Technological Development*" have been realized as planned.

The transformation capability of marine S&T achievements has shown sound progress. The transformation rate of marine S&T achievements refers to the percentage of marine S&T achievements which can realize self-transformation or be transformed into production and are at the application or production stage, and have reached the level of mature application, in the total application of marine S&T achievements. Whether marine S&T achievements can be quickly and effectively put into practical productivity or not will determine the economic development and growth of a country. Accelerating the transformation of marine S&T achievements into practical productivity, and promoting the upgrading and application of new technologies and new products, are a key link in the progress of marine science and technology. It is also the key to the transformation of the marine economy from extensive development to intensive development. *"National '12th Five-Year Plan' for Marine Economy Development"* put forward that "more than 50% of the transformation rate of marine S&T achievements in 2015 will have been realized." According to the marine S&T statistics provided by the Ministry of Science and Technology and the registered data about marine S&T achievements, the transformation rate of marine S&T achievements from 2000 to 2015 has reached 50.4% (See Appendix 7 for calculation process), smoothly realizing the objectives of the "*12th Five Year*" planning.

III. Evaluation and Analysis of National Marine Innovation Index

National marine innovation index is a composite index, which consists of five sub-indexes: marine innovation resources, marine knowledge creation, marine enterprise innovation, marine innovation performance and marine innovation environment. According to the availability of basic data, this report selects 25 indexes (See Appendix 1 for the indicator system) which can comprehensively reflect marine innovation activities to provide a picture of the quality, efficiency and capability of marine innovation.

National marine innovation index has been rising notably and marine innovation capability has been increasing significantly. If the base value of national marine innovation index in 2001 was set at 100, then the national marine innovation index in 2015 was 917. The average annual growth rate of national marine innovation index during the 2001-2015 period was 21.90%.

The sub-index of marine innovation resources has been on the rise, with average annual growth rate of 6.92% from the period of 2001 to 2015, among which the average annual growth rate of the indicators "R&D fund input intensity" and "R&D personnel input intensity" were 15.34% and 11.49% respectively, becoming the main force which stimulated the rising of the sub-index of marine innovation resources.

The sub-index of marine knowledge creation has maintained a robust increase with an average annual growth rate of 21.80%; the indicators of "number of invention patent applications per 100 million US dollars of economic output" and "number of invention patent grants per 10 thousand R&D personnel" have been increasing relatively faster with their average annual growth rate of 30.45% and 32.84% respectively, which are higher than other index values, becoming the leading factors in boosting the increase of marine knowledge creation.

The sub-index of marine enterprise innovation shows the most rapid increasing trend among these 5 sub-indexes, with the average annual growth rate of 64.23%, mainly benefiting from the leap of 2009, of which, the average annual growth rate of the indicator "ratio of enterprise R&D funds to the added value of major marine industries" reached 114.10%, the primary factor stimulating the rise of this sub-index.

The sub-index of marine innovation performance has been in a relatively slow growth trend with an average annual growth rate of only 4.54% among the five sub-indexes. Among the six indicators of the innovation performance sub-index, the growth of "marine labor productivity" remains relatively steady with an average annual growth rate of 10.50%, thereby playing an active role in boosting the growth of marine innovation performance.

The sub-index of marine innovation environment has stayed on an upward trajectory for 14 consecutive years with an average annual growth rate of 10.78%. It had made a big leap forward especially during the period from 2008 to 2010 thanks to the rapid increase of the indicators of "per capita gross ocean product of coastal areas" and "number of fresh graduates of marine study with associate degree or above".

1. Comprehensive Evaluation of Marine Innovation Index

1.1 National marine innovation index has been rising notably.

If the base value of our national marine innovation index of 2001 was set at 100, then the national marine innovation index of 2015 reached a peak at 917 (See Figure 3-1). The average annual growth rate was 21.90% during the period of 2001-2015.

National marine innovation index has maintained an upward trend during the 2001-2015 period. The growth speed has experienced different degrees of fluctuations, with a crest value appearing in 2009. National marine innovation index increased from 203 in 2008 to 568 in 2009, and the growth rate reached the peak value of 179.30%, which is mainly due to the effective measures our country took to deal with the international financial crisis in 2008. Thus remarkable results have been achieved especially in marine field. With the year of 2009 as a demarcation point, national marine innovation index had maintained a steady upward trend during the 2001-2008 period with an average annual growth rate of 10.86%. While in 2009 and the years afterwards, i.e. during the 2009-2015 period, national marine innovation index has remained above 500 with an average annual growth rate of 32.95% (See Figure 3-1).

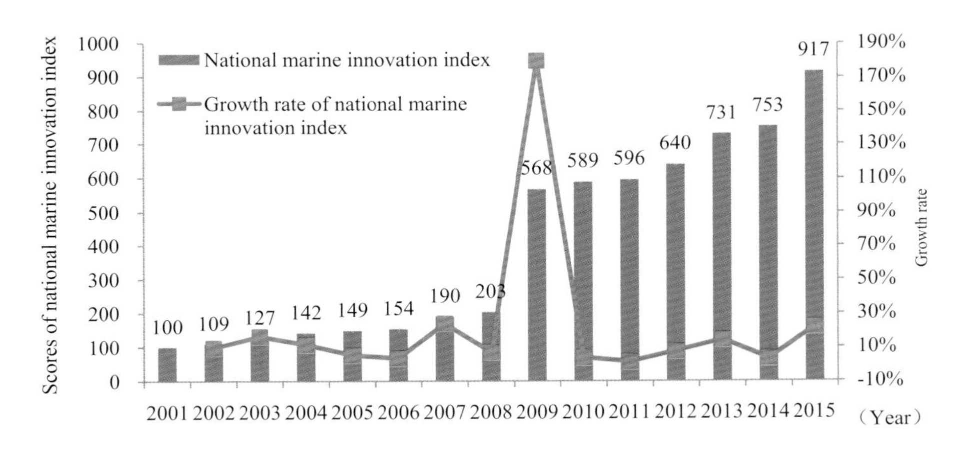

Figure 3-1 Year-on-year Variation of National Marine Innovation Index and Growth Tendency

1.2 National marine innovation index is closely related to the five sub-indexes.

The five sub-indexes have different influences on national marine innovation index, presenting different upward trends (See Table 3-1 and Figure 3-2). The sub-index of marine innovation resources and the variation trend of national marine innovation index are basically the same, with only the year of 2010 seeing a negative fluctuation in the sub-index of marine innovation resources. The scores of the sub-index of marine knowledge creation are higher than national marine innovation index with the closest trend change, reflecting comparatively greater positive contribution of marine knowledge creation to national marine innovation index. After 2009, the scores of the sub-index of marine enterprise innovation are far higher than national marine innovation index, reflecting great positive contribution of marine enterprise innovation to the growth of national marine innovation index. The maximum difference is seen between the sub-index of marine innovation performance and the variation trend of national marine innovation index. Basically its sub-index presents steady and slow linear growth and the average annual growth rate can see small-scale fluctuation which differs greatly from the growth rate of national marine innovation index. The sub-index of marine innovation environment had remained closest to the scores and trend of the national marine innovation index during the 2001-2008 period but lower than national marine innovation index during the 2009-2015 period in spite of close trend change.

Table 3-1 Variation of National Marine Innovation Index and Their Sub-indexes

Year	Comprehensive index	Sub-index				
	National marine innovation index	Marine innovation resources	Marine knowledge creation	Marine enterprise innovation	Marine innovation performance	Marine innovation environment
	A	B_1	B_2	B_3	B_4	B_5
2001	100	100	100	100	100	100
2002	109	104	126	101	108	108
2003	127	107	170	137	105	117
2004	142	108	213	156	106	126
2005	149	109	237	149	111	139
2006	154	110	239	142	120	158
2007	190	156	303	176	124	193
2008	203	162	340	173	134	207
2009	568	212	654	1530	138	305
2010	589	209	717	1535	148	335
2011	596	214	777	1467	158	363
2012	640	219	917	1518	166	379
2013	731	239	1064	1788	171	391
2014	753	239	1085	1867	179	395
2015	917	234	1319	2454	185	393

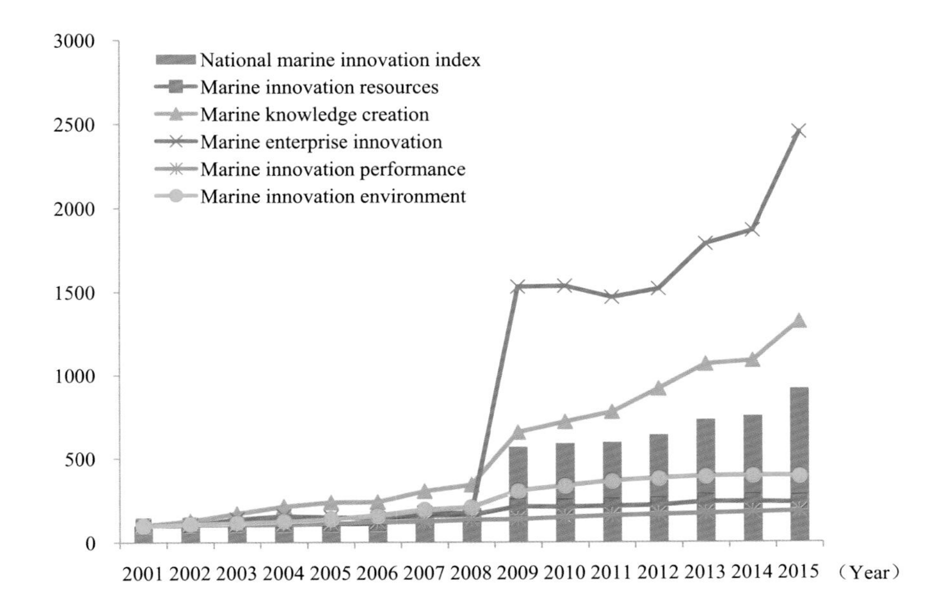

Figure 3-2 Year-on-year Variation Trend of National Marine Innovation Index and Their Sub-indexes

The sub-index of national marine innovation resources input has maintained an average annual growth rate of 6.92% during the period of 2001-2015. Except for the negative growth rate of 1.50% in 2010 and 1.96% in 2015 (See Table 3-2), the average annual growth rate has remained positive in all other years, which fully embodies the development trend of continuous increasing input in marine innovation resources of China.

Table 3-2 Growth Rate (%) of National Marine Innovation Index and Their Sub-indexes

Year	Comprehensive index	Sub-index				
	National marine innovation index	Marine innovation resources	Marine knowledge creation	Marine enterprise innovation	Marine innovation performance	Marine innovation environment
	A	B_1	B_2	B_3	B_4	B_5
2001	-	-	-	-	-	-
2002	9.45	3.95	26.33	0.89	7.59	8.50
2003	16.23	2.90	34.48	35.76	-2.44	8.13
2004	11.41	0.61	25.12	13.77	1.40	7.62
2005	5.08	1.35	11.49	-4.46	4.29	9.93
2006	3.25	0.84	0.91	-4.77	7.74	14.15

Year	Comprehensive index	Sub-index				
	National marine innovation index	Marine innovation resources	Marine knowledge creation	Marine enterprise innovation	Marine innovation performance	Marine innovation environment
	A	B_1	B_2	B_3	B_4	B_5
2007	23.81	42.24	26.68	23.80	4.07	21.61
2008	6.77	3.27	12.37	-1.19	7.92	7.30
2009	179.30	31.11	92.25	782.40	2.37	47.48
2010	3.70	-1.50	9.60	0.28	7.78	9.97
2011	1.17	2.33	8.26	-4.39	6.81	8.25
2012	7.40	2.50	18.03	3.45	5.03	4.51
2013	14.21	9.07	16.13	17.77	3.14	3.09
2014	3.07	0.14	1.95	4.45	4.37	0.99
2015	21.78	-1.96	21.60	31.43	3.55	-0.58
Average annual growth rate	21.90	6.92	21.80	64.23	4.54	10.78

The sub-index of marine knowledge creation has made greater contribution to the significant improvement of marine innovation capability with an average annual growth rate of 21.80%, remaining basically the same average annual speed as that of the national marine innovation index (See Figure 3-3). It proves that marine science and research capability is advancing rapidly, and the creation, transformation and application of marine knowledge provide a strong support for marine innovation activities. The advancement of marine knowledge creation capability provides crucial support for strengthening national original innovation capability and improving independent innovation level.

The sub-index of marine enterprise innovation contributes the most to the significant increase of marine innovation capability of our country, with an average annual growth rate reaching 64.23%, which ranks the highest among the five sub-indexes (See Figure 3-4), indicating that the innovation capability of our marine enterprises is advancing rapidly and providing crucial support for improving the level of national marine innovation.

Promoting marine economic development is the ultimate goal of carrying out marine innovation activities, which is also an indispensable factor for the evaluation of marine innovation capability. In terms of the variation trend of recent years, the marine innovation performance of China has been on a steady rise. The sub-index of the marine innovation performance of China has maintained an average annual growth rate of 4.54% during the 2001-2015 period. Except for the negative growth of 2003, the other years all see positive growth trends, and the peak value of 7.92% appeared in 2008 (See Table 3-2).

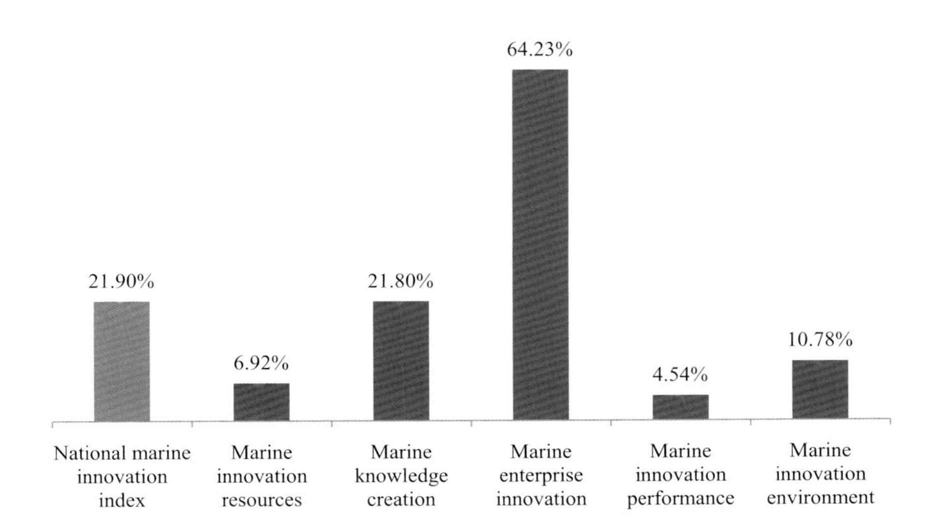

Figure 3-3 Average Annual Growth Rate of National Marine Innovation Index and Sub-indexes from 2001 to 2015

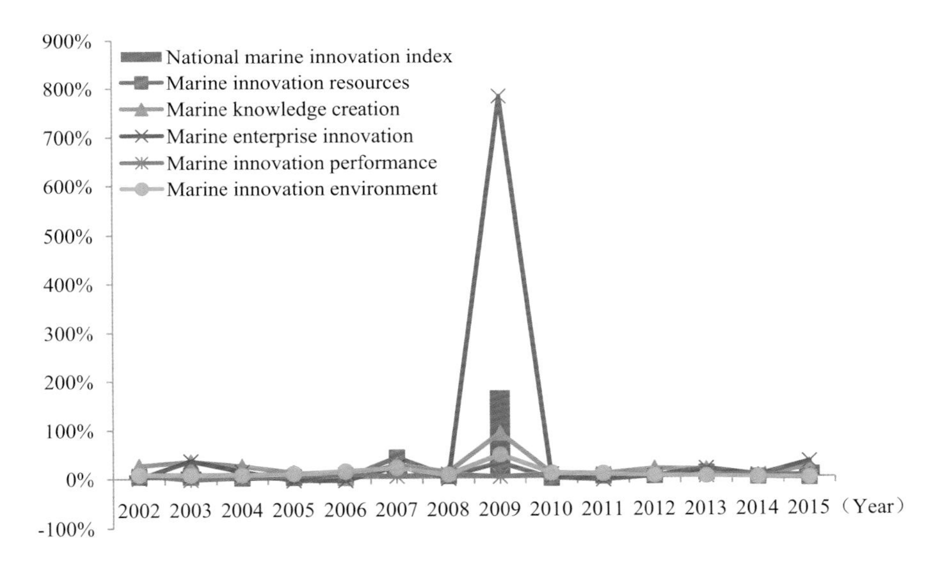

Figure 3-4 Year-on-year Growth Rates of National Marine Innovation Index and Sub-indexes

Marine innovation environment is an important guarantee for smooth carrying out marine innovation activities. Since the promulgation and implementation of *"Outline of the National '12th Five-Year Plan' for Marine Scientific and Technological Development"*, the overall environment of our country's marine innovation has seen great improvement. The sub-index of marine innovation environment has been maintaining an upward trend with an average annual growth rate of 11.66% during the period of 2001-2014 (See Table 3-2). However, the year of 2015 saw negative growth for the first time, ranking the fourth among the five sub-indexes (See Figure 3-3).

2. Evaluation of the Sub-index of Marine Innovation Resources

Marine innovation resources can reflect a country's investment in marine innovation activities. The supply capability of innovative human resources and infrastructure input levels that innovation relies on are basic guarantees for the sustainable development of innovation activities. The sub-index of marine innovation resources adopts the following five indicators: R&D funds input strength, R&D personnel input strength, proportion of staff with intermediate and senior professional titles in total S&T personnel, proportion of S&T staff in total personnel of marine scientific research institutes and the number of projects undertaken per 10 thousand scientific research personnel. The investment and configuration capabilities of our country's marine innovation resources can be evaluated from the perspectives of capital input and human resources input based on the above indicators.

2.1 The sub-index of marine resources has been rising steadily.

The sub-index of marine innovation resources in 2015 scored 234 points (See Table 3-3), with slight decrease compared to 2014. The average annual growth rate during the period of 2001-2015 was 6.92%. In terms of the historical performance, the sub-index of marine innovation resources increased most notably both in 2007 and 2009 when the average annual growth rate was 42.24% and 31.11% respectively. After 2009, the sub-index of marine innovation resources has shown an upward fluctuation on a small scale and reached a historical peak value at 239 in 2013 and 2014 followed by a slight decrease in 2015.

2.2 The change of indicators has their own characteristics.

In terms of the variation trend in the scores of the five indicators of marine innovation resources (See Figure 3-5 and Figure 3-6), two indicators increased rapidly, two remained basically unchanged and one showed a periodic upward trend as a whole. Among them, "R&D funds input strength" reveals the biggest fluctuation, followed by the indicator of "R&D personnel input strength". The above two indicators had maintained an upward trend during the 2001-2015 period, when the average annual growth rate was 15.34% and 11.49% respectively, serving as the main force that gave an impetus to the overall rise of the sub-index of marine innovation resources. However, the scores of these two indicators in 2015 were slightly lower than those of 2014.

Table 3-3 Sub-index of Marine Innovation Resources and the Scores of Its Indicators

Year	Sub-index	Indicator				
	Marine innovation resources	R&D funds input degree	R&D personnel input degree	Proportion of staff with intermediate and senior professional titles in the total S&T personnel	Proportion of S&T staff in total personnel of marine scientific research institutes	Number of projects undertaken per 10 thousand scientific research personnel
	B_1	C_1	C_2	C_3	C_4	C_5
2001	100	100	100	100	100	100
2002	104	115	94	105	99	106
2003	107	123	88	104	100	120
2004	108	109	93	107	103	127
2005	109	102	87	112	103	140
2006	110	104	85	112	104	145
2007	156	217	173	114	108	171
2008	162	229	179	111	111	178
2009	212	435	263	106	110	144
2010	209	403	264	111	114	152
2011	214	410	286	110	111	151
2012	219	417	293	113	113	159
2013	239	484	339	103	113	153
2014	239	470	338	111	115	160
2015	234	449	325	119	116	163

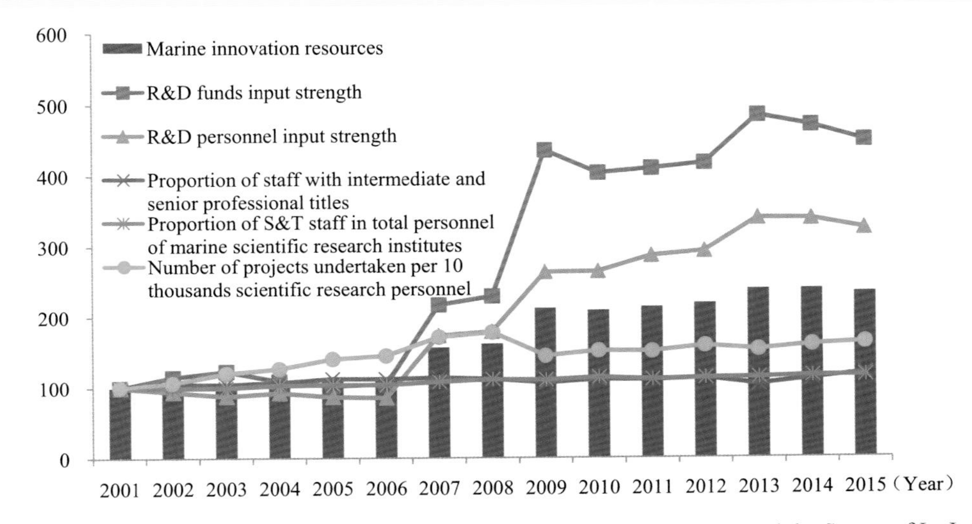

Figure 3-5 Variation Trend of the Sub-index of Marine Innovation Resources and the Scores of Its Indicators

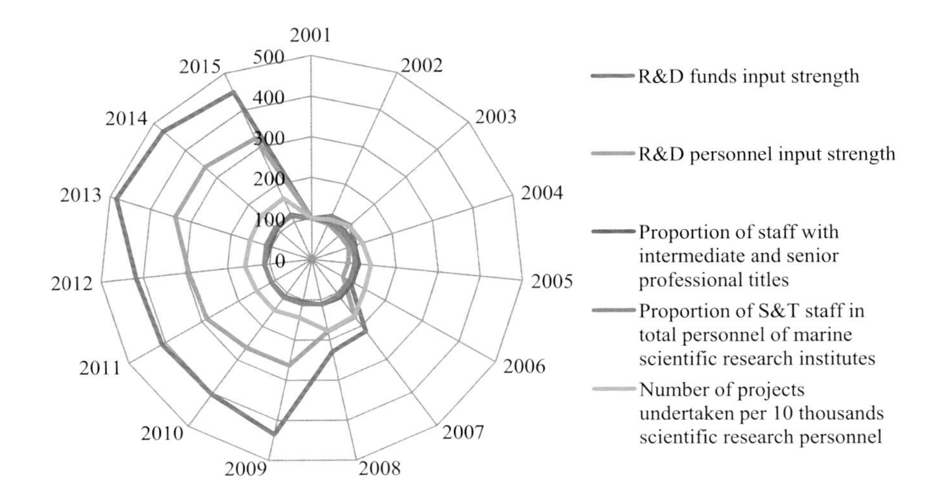

Figure 3-6 Sub-index of Marine Innovation Resources and Comparative Analysis of the Scores of Its Indicators

The indicator of "proportion of staff with intermediate and senior professional titles in total S&T personnel" reflects the strength of top talented personnel engaged in marine S&T activities of a country, while the indicator of "proportion of S&T staff in total personnel of marine scientific research institutes" reflects the degree of scientific research strength of a country's marine innovation activities. Since 2001, the growth rate of the two indicators has basically stayed at the same level, and their average annual growth rates during the 2001-2015 period were 1.35 % and 1.06% respectively, showing a slow but stable trend.

The indicator of "the number of projects undertaken per 10 thousand scientific research personnel" can reflect the intensity of S&T personnel engaged in marine innovation activities. With the year of 2009 as the demarcation point, it had maintained a steady upward trend during the 2001-2008 period with an average annual growth rate of 8.67%. In 2009, it showed a negative growth. After 2009, it had risen steadily until 2014 when it scored relatively high but still lower than that of 2008. The average annual growth rate was 3.90% during the 2010-2015 period.

3. Evaluation of the Sub-index of Marine Knowledge Creation

Marine knowledge creation is the direct output of innovation activities, which can reflect the capabilities of scientific research output and knowledge dissemination in the marine field of a country. The sub-index of marine knowledge creation selects the following five indicators: the number of invention patent applications with economic output of 100 million US dollars, the number of invention patent grants per 10 thousand R&D personnel, S&T works published in the current year, the number of S&T papers published per 10 thousand scientific research personnel and proportion of papers published abroad in total articles. To demonstrate the capability and level of our country's marine knowledge creation by means of the above indicators can not only reflect the output effect of S&T achievements, but also comprehensively take the output of various achievements such as invention patents, S&T papers and S&T works, etc. into account.

3.1 The sub-index of marine knowledge creation has been increasing rapidly.

In terms of the sub-index of marine knowledge creation and its growth rate (See Table 3-4 and Figure 3-7), the sub-index of our country has been advancing significantly from 100 in 2001 to 1319 in 2015 with average annual growth rate standing at 21.80%. It can be seen from Figure 3-7 that the growth of the marine knowledge creation sub-index can be approximately divided into two stages. With the year 2008 as the demarcation point, the first stage refers to the years before 2008. In the first stage, marine knowledge creation presented a relatively slow upward trend and it was a stage of low-speed growth with an average annual growth rate of 19.63%. The second stage is after 2008. During the second stage, a stage of high-speed growth, the sub-index of the marine knowledge creation increased rapidly, with its average annual growth rate reaching 23.97% during the 2008-2015 period and peaking at 92.25% in 2009.

Table 3-4 Sub-index of Marine Knowledge Creation and the Scores of Its Indicators

Year	Sub-index	Indicator				
	Marine knowledge creation	Number of invention patent applications per 100 million US dollars of economic output	Number of invention patent grants per 10 thousand R&D personnel	S&T works published in the current year	Number of S&T papers published per 10 thousand scientific research personnel	Proportion of papers published abroad as total articles
	B_2	C_6	C_7	C_8	C_9	C_{10}
2001	100	100	100	100	100	100
2002	126	156	133	125	115	103
2003	170	282	196	143	131	97
2004	213	278	373	160	142	108
2005	237	241	463	191	174	117
2006	239	207	534	132	184	139
2007	303	339	452	268	247	208
2008	340	414	552	291	244	202

Year	Sub-index	Indicator				
	Marine knowledge creation	Number of invention patent applications per 100 million US dollars of economic output	Number of invention patent grants per 10 thousand R&D personnel	S&T works published in the current year	Number of S&T papers published per 10 thousand scientific research personnel	Proportion of papers published abroad as total articles
	B_2	C_6	C_7	C_8	C_9	C_{10}
2009	654	1229	1139	519	208	177
2010	717	1515	1178	479	196	219
2011	777	1477	1461	528	201	215
2012	917	1539	1958	651	210	225
2013	1064	1920	2144	768	198	292
2014	1085	1729	2547	651	196	302
2015	1319	1654	3734	711	186	312

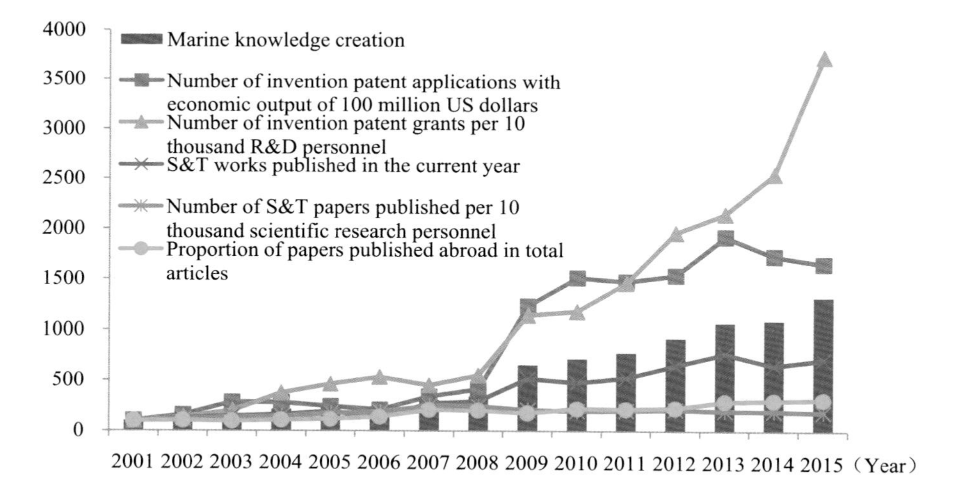

Figure3-7 Variation Trend of the Sub-index of Marine Knowledge Creation and the Scores of Its Indicators

3.2 The indicators make different contributions.

In terms of the variation trend of the five indicators of marine knowledge creation (See Figure 3-8), the two indicators of "number of invention patent applications with economic output of 100 million US dollars" and "number of invention patent grants per 10 thousand R&D personnel" fluctuate most significantly. The above two indicators rose rapidly especially during the 2008-2009 period, increasing from 414 and 552 in 2008 to 1229 and 1139 in 2009, with their average annual growth rate reaching 197.04% and 106.50% respectively. They show slight fluctuations in other years. On the whole, the two indicators has been advancing steadily and relatively fast during the period of 2001-2015, during which the average annual growth rate reached 30.45% and 32.84% respectively. These two indicators, whose scores are much higher than other indicators, have become the impetus for boosting the rising of marine knowledge creation.

The indicator of "S&T works published in the current year" has presented a steady upward trend with the average annual growth rate of 19.10% during the 2001-2015 period. The indicator had increased slowly at an average annual growth rate of 17.59% during the period of 2001-2005 and declined slightly in 2006. The 2006-2007 period and 2008-2009 period are two periods when this indicator rose rapidly and the fastest with average annual growth rate at 102.86% and 78.57% respectively. The score of the indicator "S&T works published in the current year" has been increasing gradually after 2009, peaking at 768 in 2013 followed by slight decrease in 2014. The score was 711 in 2015.

"Number of S&T papers published per 10 thousand scientific research personnel", namely the number of S&T papers published by every 10 thousand scientific research personnel, reflects the output efficiency of scientific research, while the "proportion of papers published abroad in total articles" which refers to the proportion of papers published abroad in total papers published in a country, reflects the dissemination degree of the S&T papers to the outside world. During 2001-2015 period, the scores of the two indicators have risen relatively slowly, with an average annual growth rate at 5.23% and 9.52%.

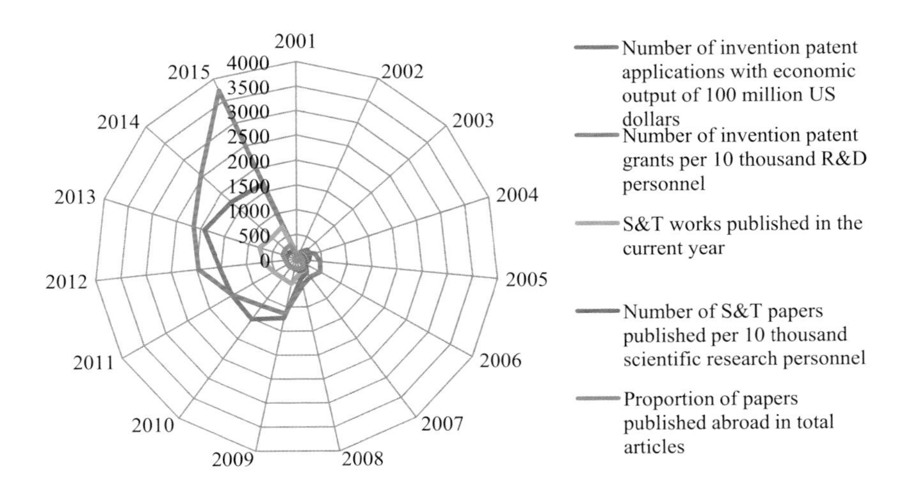

Figure 3-8 Sub-index of Marine Knowledge Creation and Comparative Analysis of Scores of Its Indicators

4. Evaluation of the Sub-index of Marine Enterprise Innovation

Enterprise is the principal place where innovation activities are carried out and also an important component of national innovation system. To a great degree, the scale and quality of marine enterprise innovation reflect the marine innovation capability and level of a country. The sub-index of marine enterprise innovation selects the following five indicators: proportion of enterprise invention patent grants in total invention patent grants, the ratio of enterprise R&D funds to the added value of major marine industries, number of invention patent grants per 10 thousand R&D personnel, independent rate of marine comprehensive technology, and proportion of enterprise R&D personnel in total R&D personnel.

4.1 The sub-index of marine enterprise innovation has been increasing rapidly.

In terms of the sub-index of marine enterprise innovation and its growth rate (See Table 3-5 and Figure 3-9), the sub-index of our country has been advancing significantly from 100 in 2001 to 2454 in 2015 with average annual growth rate reaching 64.23%, which is the fastest growth rate among the five sub-indexes. It can be seen from Figure 3-5 that the growth of marine enterprise innovation sub-index can be approximately divided into two stages. With the year 2008 as the demarcation point, the first stage refers to the years before 2008, a stage of low-speed growth when the marine enterprise innovation presented a relatively slow upward trend with an annual average growth rate of 9.11%. The second stage is after 2008, a stage of high-speed growth when the sub-index of the marine enterprise innovation increased rapidly with an average annual growth rate reaching 119.34% during the period of 2008 to 2015, peaking at 782.40% in 2009.

Table 3-5 Sub-index of Marine Enterprise Innovation and the Scores of its Indicators

Year	Sub-index	Indicator				
	Marine enterprise innovation	Proportion of enterprise invention patent grants in total invention patent grants	Ratio of enterprise R&D funds to the added value of major marine industries	Number of invention patent grants per 10 thousand R&D personnel	Independent rate of marine comprehensive technology	Proportion of enterprise R&D personnel in total R&D personnel
	B_3	C_{11}	C_{12}	C_{13}	C_{14}	C_{15}
2001	100	100	100	100	100	100
2002	101	74	128	80	99	124
2003	137	100	195	111	102	177
2004	156	81	247	160	101	189

Year	Sub-index	Indicator				
	Marine enterprise innovation	Proportion of enterprise invention patent grants in total invention patent grants	Ratio of enterprise R&D funds to the added value of major marine industries	Number of invention patent grants per 10 thousand R&D personnel	Independent rate of marine comprehensive technology	Proportion of enterprise R&D personnel in total R&D personnel
	B_3	C_{11}	C_{12}	C_{13}	C_{14}	C_{15}
2005	149	66	222	182	106	168
2006	142	70	136	245	104	153
2007	176	73	323	257	98	128
2008	173	60	325	288	80	115
2009	1530	547	5042	1563	100	399
2010	1535	553	5045	1560	98	417
2011	1467	518	4500	1804	95	419
2012	1518	519	4144	2401	101	423
2013	1788	588	4800	3034	101	415
2014	1867	612	4211	4023	103	387
2015	2454	786	2698	8336	96	352

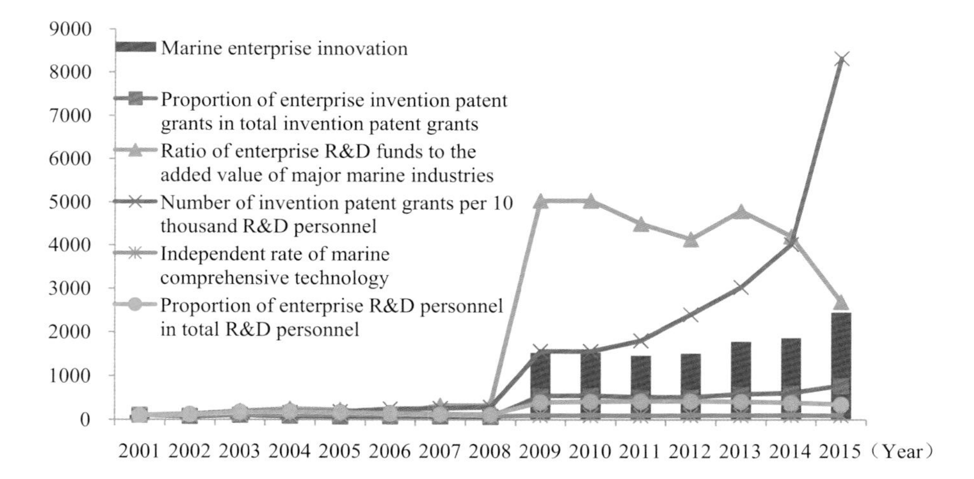

Figure 3-9 Variation Trend of the Sub-index of Marine Enterprise Innovation and the Scores of Its Indicators

4.2 The indicators make different contributions.

In terms of the variation trend of the five indicators of marine enterprise innovation (See Figure 3-6), the two indicators of "ratio of enterprise R&D funds to the added value of major marine industries" and "number of invention patent grants per 10 thousand R&D personnel" fluctuate most significantly, especially during the period from 2008 to 2009 when the above two indicators rose rapidly, increasing from 325 and 288 in 2008 to 5042 and 1563 in 2009 respectively. They showed slight fluctuations in other years. On the whole, the two indicators has presented a rapid increase trend during the period of 2001-2015, with average annual growth rate reaching 114.10% and 56.21% respectively. These two indicators, whose scores are much higher than other indicators, have become the major impetus for boosting the rising of marine enterprise innovation sub-index.

The indicator of "proportion of enterprise invention patent grants in total invention patent grants" has been increasing rapidly during the 2001 to 2015 period with an average annual growth rate of 60.86%. Like other four indicators of marine enterprise innovation, it had risen most rapidly during the period from 2008 to 2009, increasing from 60 in 2008 to 547 in 2009, which is the main factor giving an impetus to the rise of marine enterprise innovation sub-index.

"Independent rate of marine comprehensive technology" adopts the average value between the proportion of the enterprise technical revenues in total technical revenues from the regular funds revenues of marine scientific research institutes and the proportion of domesticly granted patent numbers in granted patent numbers of the current year, which can reflect technical self-sufficiency of our country's marine industry. During the period of 2001-2015, the indicator of "independent rate of marine comprehensive technology" has presented slight fluctuations with the lowest score in 2008. After that, it has increased steadily but saw a slight decrease in 2011 and increased to 103 until 2014, followed again by a slight decrease at 96 in 2015. Overall, this indicator presents a stable and slow increase trend, with an average annual growth rate of 0.06%.

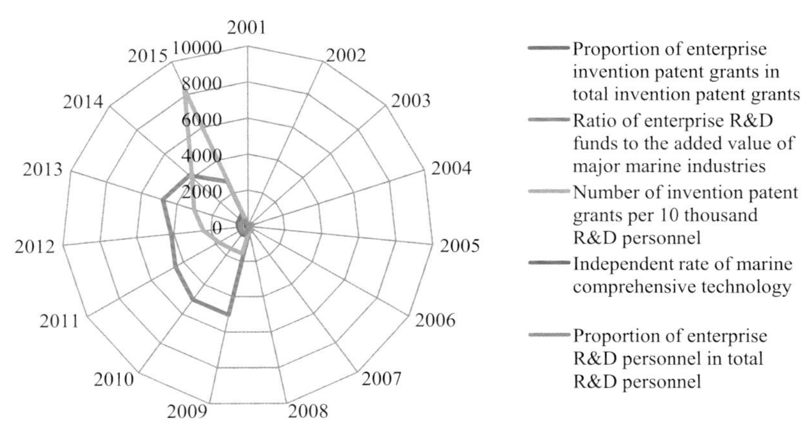

Figure 3-10 Sub-index of Marine Enterprise Creation and Comparative Analysis of the Scores of Its Indicators

During the period of 2001-2015, the indicator "proportion of enterprise R&D personnel in total R&D personnel" has risen rapidly with average annual growth rate of 18.77%.

5. Evaluation of the Sub-index of Marine Innovation Performance

The marine innovation performance can reflect the effect and impact of a country's marine innovation activities. The sub-index of marine innovation performance selects the following six indicators: transformation rate of marine S&T achievements, contribution rate of marine S&T progress, marine labor productivity, proportion of management services of science and research education in gross ocean product, marine economic output per unit energy of consumption, and the proportion of gross ocean product in gross domestic product. The effect and impact brought by the marine innovation activities of China can be reflected by means of the above indicators.

5.1 The sub-index of marine innovation performance rises sequentially.

The sub-index of marine innovation performance and year-on-year scores of the indicators can be seen from Table 3-6. In terms of the scores of the indicators, the sub-index of the marine innovation performance of China rose from 100 in 2001 to 185 in 2015, showing a steady and orderly upward trend with average annual growth rate of 4.54%, which is the slowest growth rate among the five sub-indexes.

Table 3-6 Sub-index of Marine Innovation Performance and the Scores of Its Indicators

Year	Sub-index	Indicator					
	Marine innovation performance	Transformation rate of marine S&T achievements	Contribution rate of marine S&T progress	Marine labor productivity	Proportion of management services of science and research education in gross ocean production	Marine economic output by unit energy of consumption	Proportion of gross ocean production in gross domestic product
	B_4	C_{16}	C_{17}	C_{18}	C_{19}	C_{20}	C_{21}
2001	100	100	100	100	100	100	100
2002	108	113	109	110	94	112	108
2003	105	123	94	108	101	103	101
2004	106	130	70	124	100	109	106
2005	111	137	64	141	96	118	110
2006	120	142	76	162	92	132	115
2007	124	146	74	180	91	144	111

Year	Sub-index	Indicator					
	Marine innovation performance	Transformation rate of marine S&T achievements	Contribution rate of marine S&T progress	Marine labor productivity	Proportion of management services of science and research education in gross ocean production	Marine economic output by unit energy of consumption	Proportion of gross ocean production in gross domestic product
	B_4	C_{16}	C_{17}	C_{18}	C_{19}	C_{20}	C_{21}
2008	134	150	89	204	92	161	109
2009	138	154	84	219	94	166	109
2010	148	157	80	261	85	192	114
2011	158	160	94	294	84	207	111
2012	166	162	100	319	86	219	111
2013	171	165	96	342	87	229	110
2014	179	167	101	378	92	225	110
2015	185	169	101	399	96	238	109

5.2 The change of indicators tends to be stable.

"Transformation rate of marine S&T achievements" is an important indicator which can measure the transformation level of marine science and technology into practical productive activities. It has maintained a slow upward trend during the 2001-2015 period with an average annual growth rate of 3.88%. Overall, the increase of transformation rate of marine S&T achievements in China was relatively more notable before 2010, and had remained stable after 2010 (See Figure 3-11).

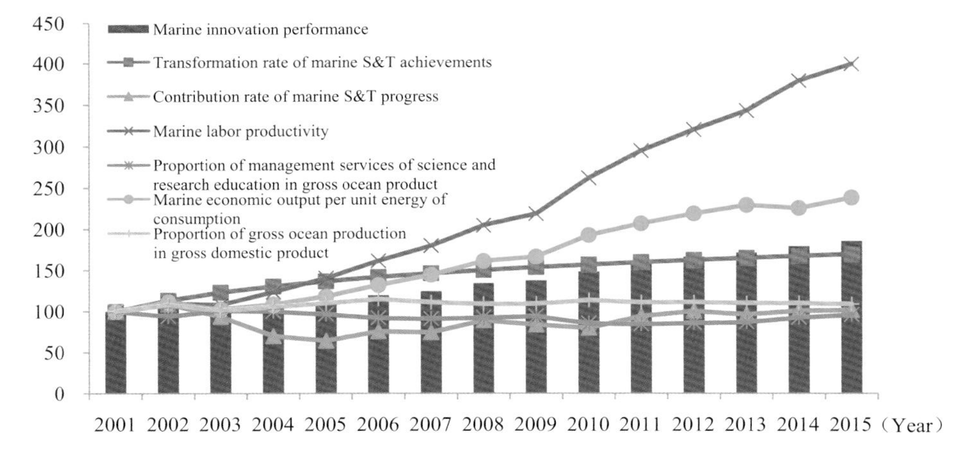

Figure 3-11 Variation Trend of the Sub-index of Marine Innovation Performance and the Scores of Its Indicators

The indicator of "contribution rate of marine S&T progress" shows slight fluctuations. During the period of 2001-2015, the contribution rate of marine S&T progress in China has maintained a steady upward trend.

"Marine labor productivity", which adopts the per capita gross ocean product of marine S&T personnel, reflects the role marine innovation activities play in marine economic output. The indicator of "marine labor productivity" has risen rapidly with an average annual growth rate of 10.50% during the period of 2001-2015, which is the one that increases most rapidly and steadily among the six indicators of the innovation performance sub-index (see Figure 3-11 and 3-12).

The indicator "proportion of management services of science and research education in gross ocean product" reflects the contribution of marine science and research, education, and management services to marine economy. Since 2001, the average annual decrease rate has been 0.21%, indicating that the contribution of marine science and research, education and management services to marine economy have shown a relative downward trend.

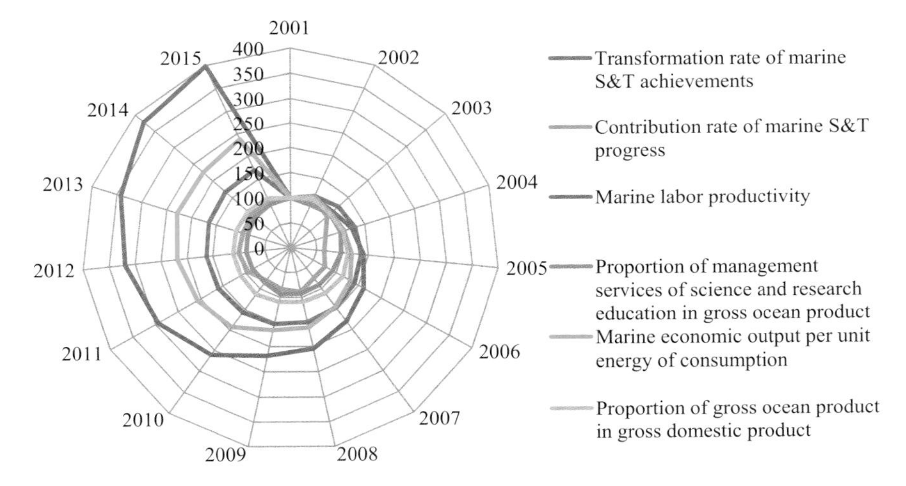

Figure 3-12 Sub-index of Marine Innovation Performance and Comparative Analysis of the Scores of Its Indicators

The indicator of "marine economic output by unit energy of consumption" refers to the gross ocean production resulting from the energy consumption of ten thousand tons of standard coal. It is used to measure the effect of energy consumption reduction brought by marine innovation and reflects the intensification level of marine economic growth of a country. It has risen rapidly with an average annual growth rate of 6.54% during the period of 2001-2015, showing a steady upward trend.

The indicator of "proportion of gross ocean product in gross domestic product" reflects the contribution of marine economy to national economy. It is used to measure the boosting role that marine innovation plays in marine economy. Figure 3-11 and Figure 3-12 show that this indicator makes no obvious change, with relatively steady scores and slow growth rate. The score of 2015 was only 9 points higher than that of 2001 and the average annual growth rate during the 2001-2014 period was only 0.66%.

6. Evaluation of the sub-index of Marine Innovation Environment

Marine innovation environment, including both hard and soft environments of innovation process, provides an important foundation and crucial guarantee for boosting the innovation capability of China. The sub-index of marine innovation environment reflects the external environment of a country that marine innovation activities rely on, which mainly refers to institutional innovation and environmental innovation. The sub-index of marine innovation environment selects the following four indicators: per capita gross ocean product of coastal areas, proportion of equipment procurement cost in R&D funds, proportion of government funds in the total science and technology funds of marine scientific research institutes and the number of fresh graduates of marine study with associate degree or above.

6.1 The marine innovation environment is improving notably.

The sub-index of marine innovation environment has been on the rise during the period of 2001-2015 (See Table 3-7, Figure 3-13 and Figure 3-14) and the score increased from 100 in 2001 to 393 in 2015 with an average annual growth rate reaching 10.786%, among which the growth rate of 2009 peaked at 47.48% mainly thanks to the rapid increase in the "number of fresh graduates of marine study with associate degree or above". The number of fresh graduates of marine study with associate degree or above in 2009 was 2.09 times that of 2008, and its score increased from 309 in 2008 to 639 in 2009. Since 2009, China's marine education has developed rapidly, giving a strong impetus to the improvement of marine innovation environment. However, in the past two years, the number of fresh graduates of marine study with associate degree or above was beginning to fall with the sub-index of marine innovation environment experiencing negative growth for the first time in 2015. More attention needs to be paid to the development of marine innovation environment.

Table 3-7 Sub-index of Marine Innovation Environment and Year-on-year Scores of Its Indicators

Year	Sub-index	Indicators			
	Marine innovation environment	Per capita gross ocean product of coastal areas	Proportion of equipment procurement cost in R&D funds	Proportion of government funds in the total science and technology funds of marine scientific research institutes	Number of fresh graduates of marine study with associate degree or above
	B_5	C_{22}	C_{23}	C_{24}	C_{25}
2001	100	100	100	100	100
2002	108	118	109	96	111
2003	117	124	108	81	157
2004	126	150	121	63	171

Year	Sub-index	Indicators			
	Marine innovation environment	Per capita gross ocean product of coastal areas	Proportion of equipment procurement cost in R&D funds	Proportion of government funds in the total science and technology funds of marine scientific research institutes	Number of fresh graduates of marine study with associate degree or above
	B_5	C_{22}	C_{23}	C_{24}	C_{25}
2005	139	180	121	64	191
2006	158	215	129	62	228
2007	193	255	180	70	267
2008	207	292	157	69	309
2009	305	314	216	51	639
2010	335	379	152	49	761
2011	363	433	123	49	847
2012	379	490	128	53	846
2013	391	528	115	55	866
2014	395	587	117	57	819
2015	393	622	115	62	772

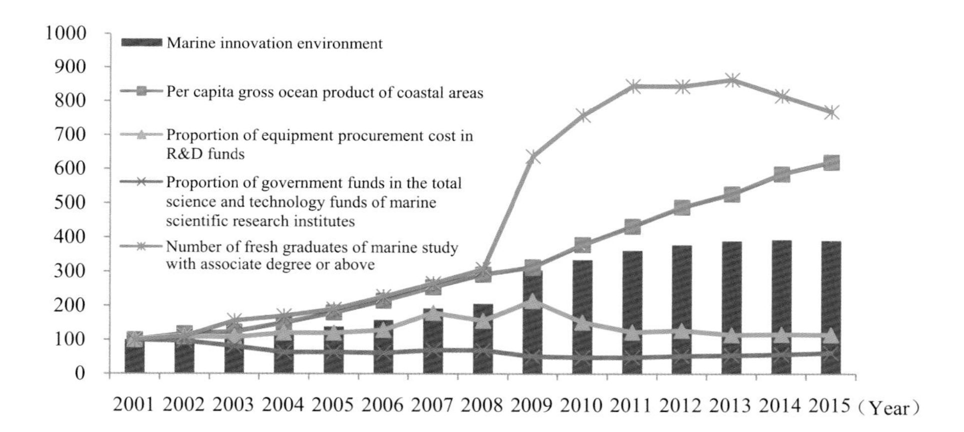

Figure 3-13 Variation Trend of the Sub-index of Marine Innovation Environment and Scores of Its Indicators

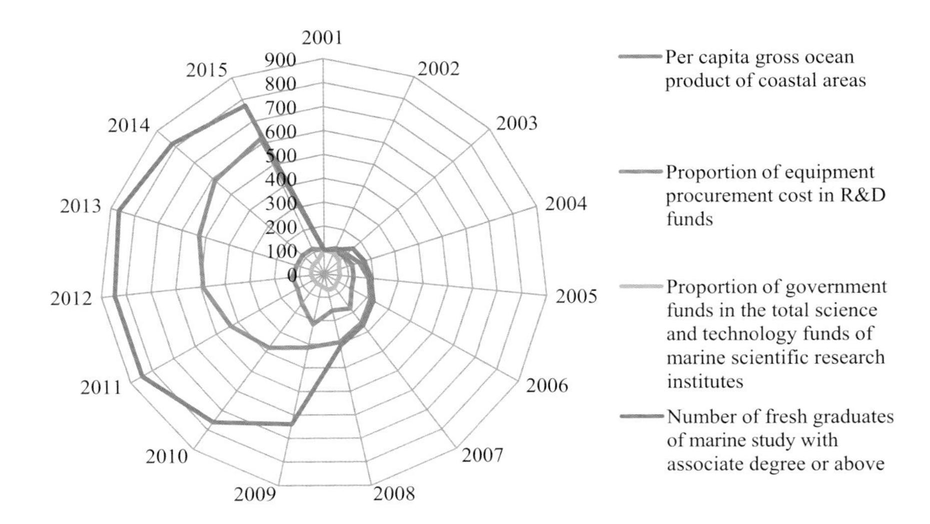

Figure 3-14 Sub-index of Marine Innovation Environment and Comparative Analysis of the Scores of Its Indicators

6.2 China has a mixed performance in the indicators of marine innovation environment.

Among the indicators of marine innovation environment sub-index, the indicators which have maintained an upward trend are "per capita gross ocean product of coastal areas" and "number of fresh graduates of marine study with associate degree or above". The score of "per capita gross ocean product of coastal areas" witnessed a notable upward trend at an average annual growth rate of 14.08% and both its score and trend remained closest to the marine innovation environment sub-index among the four indicators. In terms of the "number of fresh graduates of marine study with associate degree or above", the indicator score in 2015 was 7.72 times that of 2001 and the average annual growth rate stood at 18.81%, which is the fastest growth rate among the four indicators.

The poorer performing indicators include "proportion of equipment procurement cost in R&D funds" and "proportion of government funds in the total science and technology funds of marine scientific research institutes". The score of "proportion of equipment procurement cost in R&D funds" showed some fluctuations with an overall declining trend, peaking in 2009 then dropping gradually afterwards from 216 in 2009 to 115 in 2015. The score of "proportion of government funds in the total science and technology funds of marine scientific research institutes" has maintained an overall downward trend too, declining from 100 in 2001 to 62 in 2015.

IV. Evaluation and Analysis of Regional Marine Innovation Index

Regional marine innovation is an important component of national marine innovation and its development will influence the pattern of national marine innovation. This report analyzes the development and characteristics of regional marine innovation and provides data foundation and decision basis for the optimization of innovation pattern in China.

It states clearly in "*Vision and Actions on Jointly Building Silk Road Economic Belt and 21st-Century Maritime Silk Road*" that "we should leverage the advantages of high degree of openness, robust economic strengths and strong catalytic role of the Yangtze River Delta, Pearl River Delta, west coast of the Taiwan Straits, Bohai Rim". Analyzing from the development concepts of "*The Belt and Road Initiative*" and looking from the perspective of our country's coastal regions, we should actively optimize the general marine economic layout, make use of complementary advantages, practice joint exploitation and play the leading role of the five economic zones—Bohai Rim, Yangtze River Delta, west coast of the Taiwan Straits, Pearl River Delta and Beibu Gulf Rim economic zones[1][2](See Appendix 9 for the Definition of Marine Economic Zone) to boost the formation of three economic circles in the north, east and south[3]of China (See Appendix 9 for the Definition of Marine Economic Circle).

In terms of the regional marine innovation index of our country's coastal provinces (cities) (See Appendix 4 for Evaluation Methods of Regional Marine Innovation Index), the eleven coastal provinces (cities) of China could be divided into four tiers in 2015. Shanghai belongs to the first tier, followed by Shandong, Guangdong and Tianjin as the second tier, Fujian, Jiangsu, Liaoning, Zhejiang and Hebei as the third tier, and Hainan and Guangxi as the fourth tier.

In terms of the regional marine innovation index of the five economic zones, the regions whose innovation was relatively strong in the coastal areas are mainly concentrated in the Pearl River Delta economic zone, Yangtze River Delta economic zone and large parts of the Bohai Rim economic zone in 2015. These areas all have their own regional innovation centers and present a polycentric development pattern.

In terms of the regional marine innovation index of the three marine economic circles, the marine economic circles of our country in 2015 presented the following characteristics: the north and east circles

---

1 This evaluation only includes 11 coastal provinces (cities) of the mainland of China, excluding Hong Kong, Macao and Taiwan.

2 The coastal provinces (cities) of the Bohai Rim economic zone which are incorporated into the evaluation are Liaoning, Hebei, Shandong, and Tianjin; the coastal provinces (cities) of Yangtze River Delta economic zone which are incorporated into the evaluation are Jiangsu, Shanghai, Zhejiang; the coastal province (city) of the west coast of the Taiwan Straits economic zone which is incorporated into the evaluation is Fujian; the coastal province (city) of the Pearl River Delta economic zone which is incorporated into the evaluation is Guangdong and the coastal provinces (cities) of Beibu Gulf Rim economic zone which are incorporated into the evaluation are Guangxi and Hainan.

3 *The "12th Five-Year" Plan for National Marine Economic Development* is the division basis of the marine economic circle. The northern marine economic circle is made up of the coasts and waters of Liaodong Peninsula, Bohai Bay and Shandong Peninsula. The coastal provinces (cities) which are incorporated into the evaluation include Tianjin, Hebei, Liaoning and Shandong. The eastern marine economic circle is composed of the coasts and waters of Jiangsu, Shanghai, and Zhejiang. The coastal provinces (cities) which are incorporated into the evaluation include Jiangsu, Zhejiang and Shanghai. The southern marine economic circle consists of the coasts and waters of Fujian, the Pearl River Estuary and its two wings, Beibu Gulf and Hainan Island. The coastal provinces (cities) which are incorporated into the evaluation include Fujian, Guangdong, Guangxi, and Hainan.

are relatively strong while the south circle is relatively weak. The regional marine innovation indexes of the northern and eastern economic circles are higher, reflecting strong original innovation capability and fully manifesting the advantages of marine talent agglomeration and the key developing areas of marine economic industries in China.

1. China's Regional Marine Innovation Development Viewed from the Perspective of Coastal Provinces (Cities)

1.1 Regional marine innovation shows clear tier characteristics.

On the whole, the eleven coastal provinces (cities) of mainland China can be divided into four tiers (See Figure 4-2) based on the scores (See Table 4-1 and Figure 4-1) of regional marine innovation index in 2015.

Table 4-1 Scores of Regional Innovation Index and Sub-indexes of Coastal Provinces (cities) in 2015

Coastal provinces (cities)	Comprehensive index	Sub-index			
	Regional marine innovation index	Marine innovation resources	Marine knowledge creation	Marine innovation performance	Marine innovation environment
	a	b_1	b_2	b_3	b_4
Shanghai	64.27	76.03	35.32	86.80	58.95
Shandong	55.60	50.79	54.41	50.99	66.20
Guangdong	52.15	50.18	61.08	53.76	43.57
Tianjin	51.92	60.61	25.50	73.67	47.90
Fujian	46.39	44.39	25.10	51.45	64.63
Jiangsu	45.09	61.92	39.15	42.10	37.22
Liaoning	44.96	55.32	60.30	31.02	33.18
Zhejiang	35.41	43.78	21.77	34.54	41.54
Hebei	32.31	46.81	37.31	14.99	30.14
Hainan	26.94	15.41	6.41	53.68	32.29
Guangxi	17.14	21.96	8.42	8.61	29.57

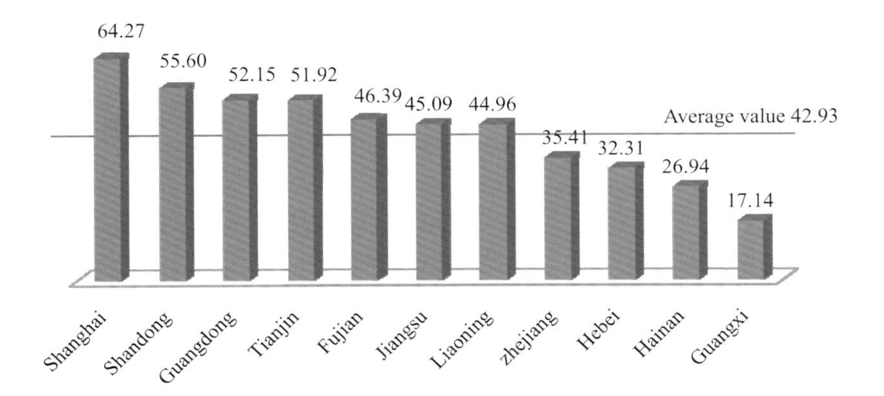

Figure 4-1 Scores of Regional Marine Innovation Index in Coastal Provinces (cities) and Average Value in 2015

In terms of regional marine innovation index, Shanghai belongs to the first tier, whose score was 64.27, equivalent to 1.50 times that of national average, and ranked top among the eleven coastal provinces (cities) of mainland China. Shanghai boasts a solid foundation for marine innovation development and shows strong original innovation capability for marine science and technology. The second tier includes Shandong, Guangdong and Tianjin, and their scores were 55.60, 52.15 and 51.92 points respectively, higher than 42.93, the average score of the eleven coastal provinces (cities). These areas have strong marine innovation foundations and have accumulated large amounts of innovation resources over time. Additionally, they possess better innovation environments and have made remarkable achievements in innovation performance. Fujian, Jiangsu, Liaoning, Zhejiang and Hebei belong to the third tier, and their scores were 46.39, 42.09, 44.96, 35.41 and 32.31 respectively, close to the average score. These areas have been developing relatively rapidly in recent years with constantly increasing innovation resources and constantly improving innovation environments. And, at the same time, knowledge creation and innovation performance are both making greater progress. The fourth tier includes Hainan and Guangxi, and their scores were 26.94 and 17.14, far below the national average. Compared horizontally, their marine innovation resources are weak, knowledge creation efficiency is not high and innovation environments are in need of improvement.

In terms of the sub-index of marine innovation resources, in 2015, the coastal provinces (cities) whose scores were above the average score are Shanghai, Jiangsu, Tianjin, Liaoning, Shandong, and Guangdong (See Figure 4-3). Among them, Shanghai scored 76.03, far higher than other areas. Shanghai ranked first among eleven coastal provinces (cities) both in the input strength of funds and human resources. Jiangsu, Tianjin, Liaoning, Shandong and Guangdong scored 61.92, 60.61, 55.32, 50.79 and 50.18 respectively mainly thanks to the high-quality marine innovation talent and human resources input.

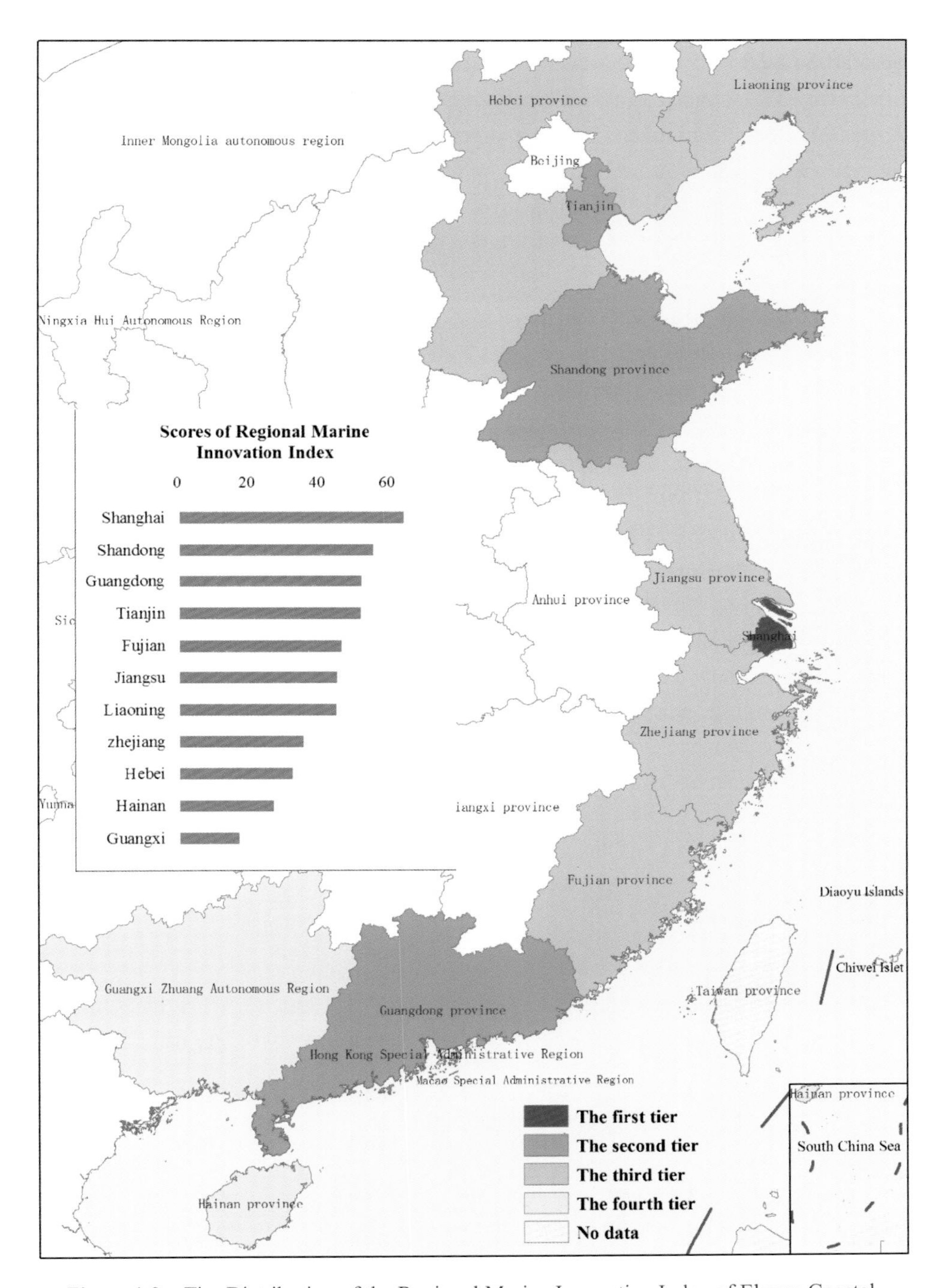

Figure 4-2 Tier Distribution of the Regional Marine Innovation Index of Eleven Coastal Provinces (cities) in 2015

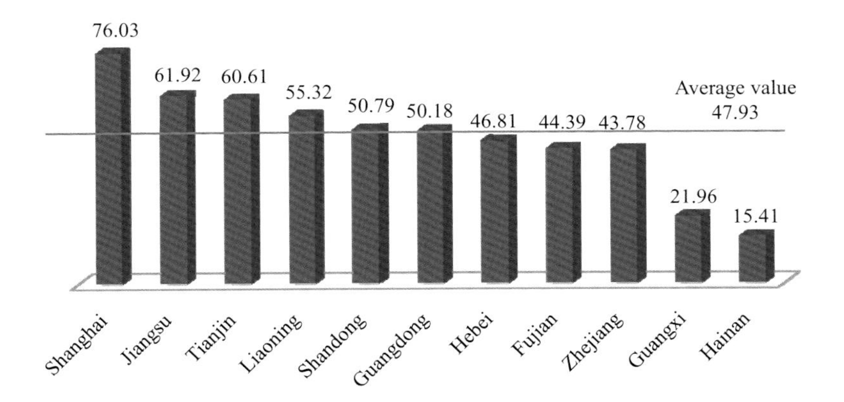

Figure 4-3 Sub-index Scores of Regional Marine Innovation Resources of Coastal Provinces (cities) and Average Score in 2015

In terms of the sub-index of marine knowledge creation, in 2015 the coastal provinces (cities) whose scores surpassed the average score are Guangdong, Liaoning, Shandong, Jiangsu, Hebei and Shanghai (See Figure 4-4). Among them, Guangdong scored 61.08 which correlated closely with its high-output, high-quality marine S&T works and papers. Liaoning scored 60.30, far higher than other areas, which mainly benefited from the marine S&T invention patents. Shandong scored 54.41 mainly thanks to marine S&T works and papers. Jiangsu scored 39.15, whose major contribution came from high-output, high-quality papers. Hebei scored 37.31 and ranked top among eleven coastal provinces (cities) in the number of publications of marine S&T works and papers, but the number of patents and the quality of the papers are in need of improvement. Shanghai's score was relatively low at only 35.32.

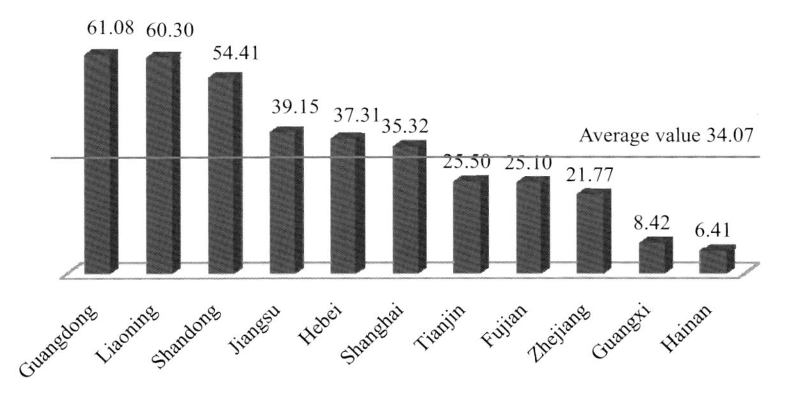

Figure 4-4 Sub-index Scores of Regional Marine Knowledge Creation of Coastal Provinces (cities) and Average Score in 2015

In terms of the sub-index of marine innovation performance, in 2015 the coastal provinces (cities) whose scores exceeded the average score are Shanghai, Tianjin, Guangdong, Hainan, Fujian and Shandong (See Figure 4-5). Among them, Shanghai scored 86.80, mainly due to the fact that it not only has far higher labor productivity than other areas, but also boasts excellent marine economic output. Benefiting from its marine economic output, Tianjin scored 73.67, just below Shanghai. Guangdong, Hainan, Fujian and Shandong scored 53.76, 53.68, 51.45 and 50.99 respectively, showing that they performed well in all aspects of marine innovation performance and their overall level is above the national average.

In terms of the sub-index of marine innovation environment, in 2015 the coastal provinces (cities) whose scores surpassed the average score are Shandong, Fujian, Shanghai and Tianjin (See Figure 4-6). Among them, Shandong scored 66.20, which is higher than other areas for possessing better marine innovation talent and government funding environments. Benefiting from state-of-the-art marine equipment and favorable government funding environment, Fujian scored 64.63. Shanghai scored 58.95 because of favorable marine innovation funding environment and higher per capita gross ocean product. Tianjin, which has high per capita gross ocean production, scored 47.90.

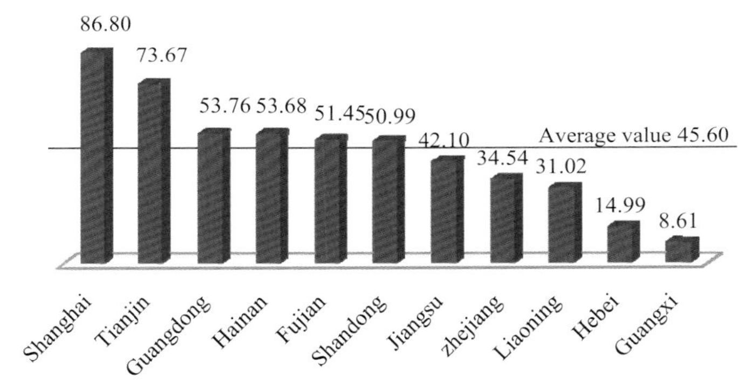

Figure 4-5 Sub-index Scores of Regional Marine Innovation Performance of the Coastal Provinces (cities) and Average Score in 2015

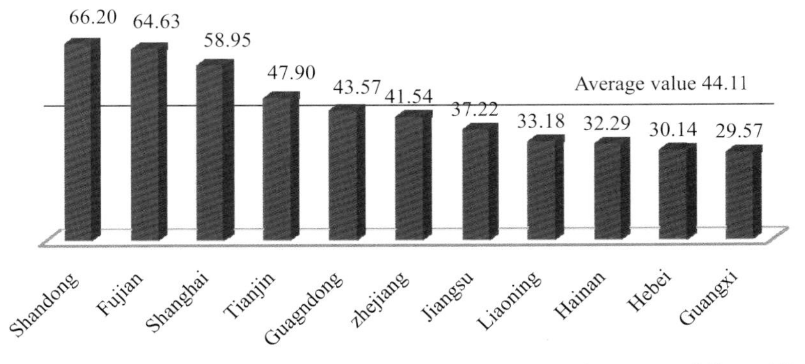

Figure 4-6 Sub-index Scores of Regional Marine Innovation Environment of Coastal Provinces (cities) and Average Score in 2015

1.2 Marine innovation capability correlates strongly with the level of economic development.

Regional marine innovation capability and the level of economic development are closely linked. As can be seen from Figure 4-7 which depicts the relationship between "per capita GDP of coastal areas" which reflects the level of economic development and "regional marine innovation index", the per capita GDP of coastal areas in the first quadrant is relatively high, the regional marine innovation index is higher than the national average, and cities in this quadrant belong to the first and second tier regions; the per capita GDP in the fourth quadrant is relatively high, but the regional marine innovation index is lower than the national average, cities in this quadrant belong to the third tier region; the per capita GDP in the third quadrant is relatively low and the regional marine innovation index is lower than the national average, Hainan and Guangxi in this quadrant belong to the fourth tier region; there is no region in the second quadrant featuring a low per capita GDP but higher regional marine innovation index than the national average. All the above show that there is strong correlation between marine innovation activities and the level of economic development of coastal areas.

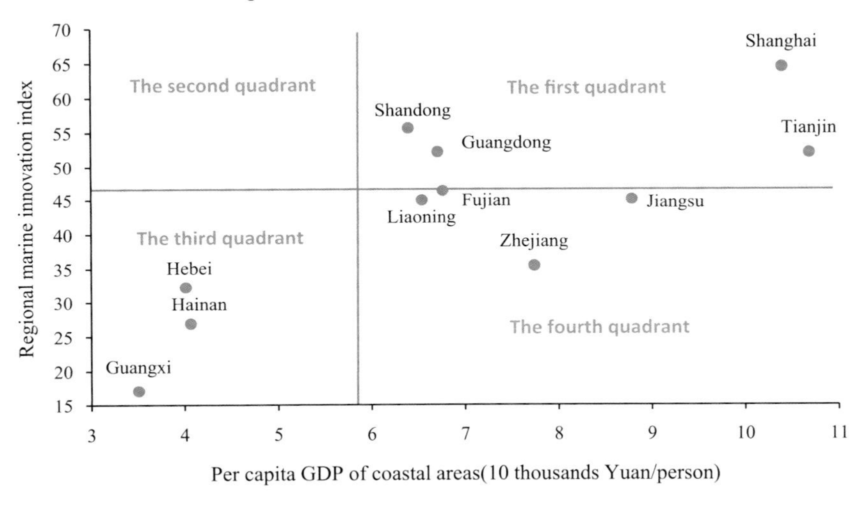

Figure 4-7 Per Capita GDP and Regional Marine Innovation Index of Coastal Provinces (cities, districts) in 2015

At the same time, there also exists strong correlation between regional marine innovation capability and the level of marine economic development. As can be seen from Figure 4-8 which depicts the relationship between "per capita GOP of coastal areas" which reflects the level of economic development and "regional marine innovation index", the per capita GOP of coastal areas in the first quadrant is relatively high, the regional marine innovation index is higher than the national average, and cities in this quadrant belong to the first and second tier regions; the per capita GOP in the fourth quadrant is relatively high, but the regional marine innovation index is close or lower than the national average, and cities in this quadrant include regions belonging to the third tier regions excluding Hebei and Henan of the fourth tier region; both the per capita

GOP and the regional marine innovation index in the third quadrant are lower than the national average, Hebei and Guangxi in this quadrant belong to the third tier and fourth tier regions; there is no region in the second quadrant featuring a low per capita GOP but higher regional marine innovation index than the national average. All the above show that there is a strong correlation between marine innovation activities and the level of marine economic development of coastal areas.

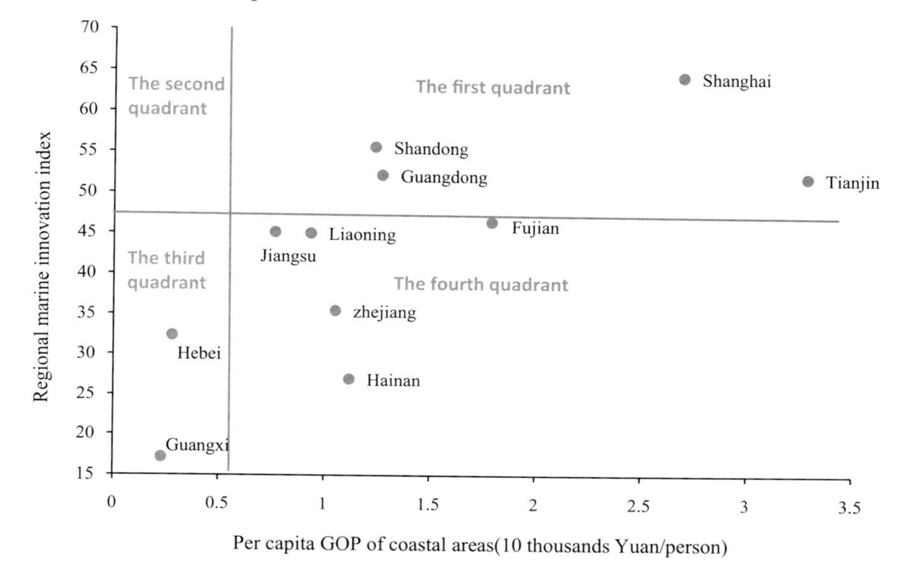

Figure 4-8 Per Capita GOP and Regional Marine Innovation Index of Coastal Provinces (cities, districts) in 2015

2. China's Regional Marine Innovation Development Viewed from the Perspective of Five Economic Zones

Detailed analysis about the Bohai Rim economic zone, Yangtze River Delta economic zone, west coast of the Taiwan Straits economic zone, Pearl River Delta economic zone and Beibu Gulf Rim economic zone is as follows:

The Bohai Rim economic zone refers to the vast economic areas consisting of the coastal areas surrounding the whole Bohai Sea and part of the Yellow Sea. It is the "Gold Coast" in the east of China which possesses perfect industrial base, abundant natural resources, strong S&T force and convenient transportation. It is the key strategic area driving the development of China's central and western regions. Consequently it plays a decisive role in the country's national economic development pattern. In 2015, its regional marine innovation index scored 46.20 (See Table 4-2 and Figure 4-9), slightly higher than the average level of the eleven coastal provinces (cities), but its regional marine innovation environment and innovation performance scored below the average. Overall, it still has room to improve its marine innovation development.

Table 4-2 Regional Marine Innovation Index and Sub-indexes of China's Five Economic Zones in 2015

Economic zone	Comprehensive index	Sub-index			
	Regional marine innovation index	Marine innovation resources	Marine knowledge creation	Marine innovation performance	Marine innovation environment
	A	b_1	b_2	b_3	b_4
Bohai Rim economic zone	46.20	44.35	53.38	44.38	42.67
Yangtze River Delta economic zone	48.26	45.90	60.57	32.08	54.48
West coast of the Taiwan Straits economic zone	46.39	64.63	44.39	25.10	51.45
Pearl River Delta economic zone	52.15	43.57	50.18	61.08	53.76
Beibu Gulf Rim economic zone	22.04	30.93	18.68	7.41	31.14
Average values	43.01	45.88	45.44	34.01	46.70

The Yangtze River Delta economic zone is located at the intersection of the coastal and riverside regions in the east of China with prominent geographical advantages and strong economic strength. With Shanghai as its core and technology-based industry as the main industry, the Yangtze River Delta economic zone is a region with significant development potential, characterized by strong technical force, bright prospects, high level of governmental support, superior environment, good educational development and sufficient talent resources. In 2015, its regional marine innovation index scored 48.26, higher than the average level of the eleven coastal provinces (cities). Vast amounts of marine innovation resources provide favorable conditions for marine S&T and economic development of the Yangtze River Delta economic zone with remarkable achievements in marine innovation.

The west coast of the Taiwan Straits economic zone has Fujian as its center and includes the neighboring areas. It borders the Pearl River Delta economic zone and the Yangtze River Delta economic zone to the south and north respectively and connects with Taiwan island and Jiangxi's hinterland to the east and west. It is a regional economic complex with unique advantages and will take the lead in making the national economy to go global. In 2015, its regional marine innovation index scored 46.39, slightly higher than the average of the eleven coastal provinces (cities). Its scores of regional marine innovation environment and performance were both above the average with a relatively great development potential, but the innovation resources and knowledge creation level were relatively low. Enterprise innovation capability was weak and marine innovation development capability is yet to be improved.

The Pearl River Delta economic zone mainly refers to Guangdong province in the south of the Chinese mainland, bordering Hong Kong and Macao Special Administrative Regions. It boasts exceptional S&T strength, abundant talent resources and rich marine resources. And it is one of the regions which has experienced the fastest economic development. Its regional marine innovation index scored 52.15, which was higher than the average level of the eleven coastal provinces (cities) and made it rank the first among the five economic zones. It has intensive marine innovation resources with fruitful knowledge creation and productive innovation performance.

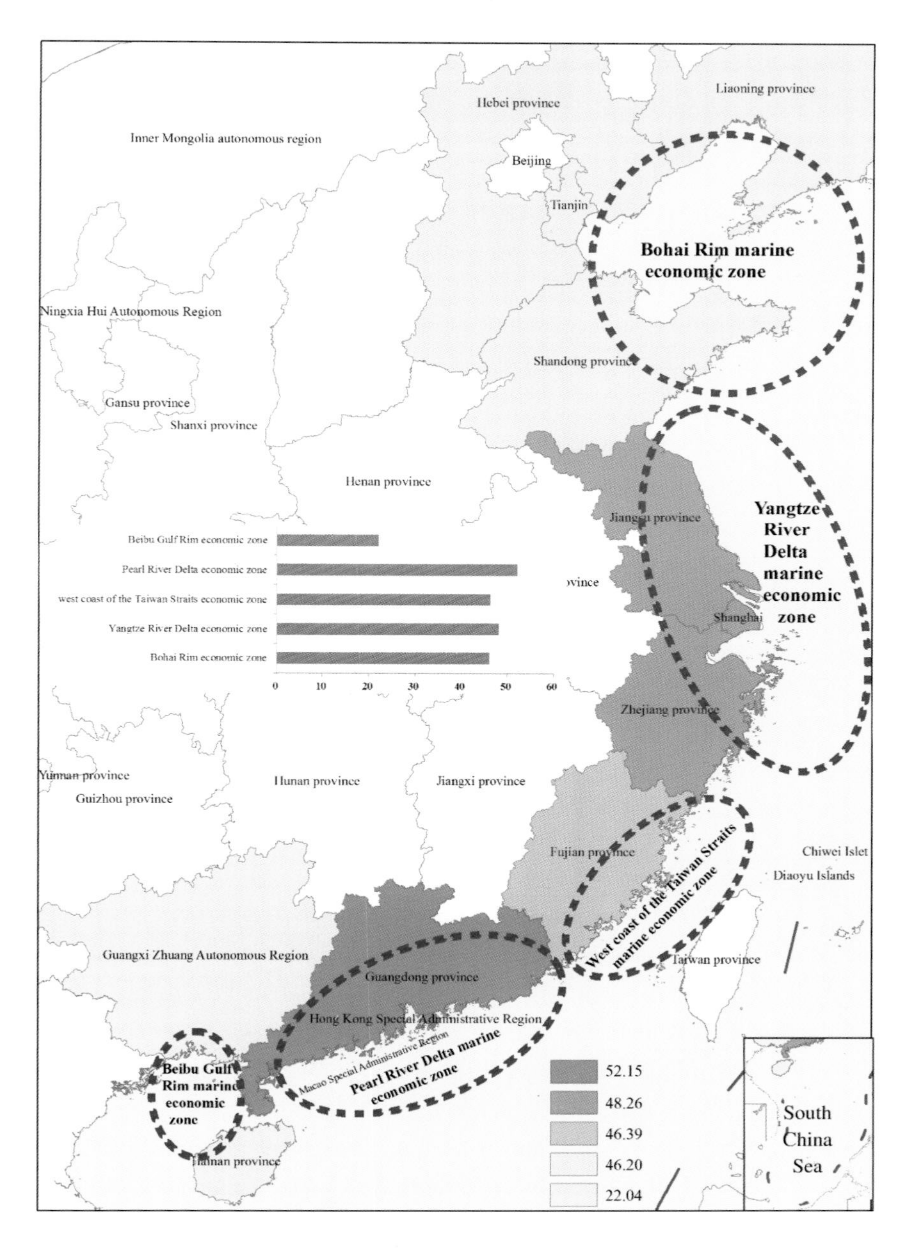

Figure 4-9 Regional Marine Innovation Index of China's Five Economic Zones in 2015

The Beibu Gulf Rim economic zone, which is the only coastal area of the western development region, is located at the intersection of South China's economic circle, southwest economic circle and ASEAN economic circle. It is the region in China connected to the ASEAN countries with both seaway and land borders and has significant geographical advantage and an outstanding strategic position. Its regional marine innovation index was only 22.05, far lower than the average level of the eleven coastal provinces (cities) and ranking at the bottom among the five economic zones. Its four sub-indexes of the innovation index were all relatively low, forming a wide gap between the Yangtze River Delta and the Pearl River Delta economic zones.

3. China's Regional Marine Innovation Development Viewed from the Perspective of Three Marine Economic Circles

The Eastern Marine Economic Circle scored 48.26 in the marine innovation index in 2015, which ranked the top among the three marine economic circles (See Table 4-3 and Figure 4-10). Among the four sub-indexes, the scores of the sub-indexes of marine innovation resources and performance were relatively higher, which stood at 60.57 and 54.48 respectively, making greater positive contributions to the marine innovation index of this region and fully demonstrating favorable conditions for regional marine S&T and economic development provided by the outstanding geographical advantage, strong economic power and high-quality marine innovation resources. The scores of marine knowledge creation and marine innovation environment were relatively lower, which were 32.08 and 45.90, thereby contributing negatively to the marine innovation index of this region (See Figure 4-11).

Table 4-3 Regional Marine Innovation Index and Sub-indexes of China's Three Marine Economic Circles in 2015

Economic circle	Comprehensive index	Sub-index			
	Regional marine innovation index	Marine innovation resources	Marine knowledge creation	Marine innovation performance	Marine innovation environment
	A	b_1	b_2	b_3	b_4
Northern marine economic circle	46.20	53.38	44.38	42.67	44.35
Eastern marine economic circle	48.26	60.57	32.08	54.48	45.90
Southern marine economic circle	35.66	32.98	25.25	41.87	42.52

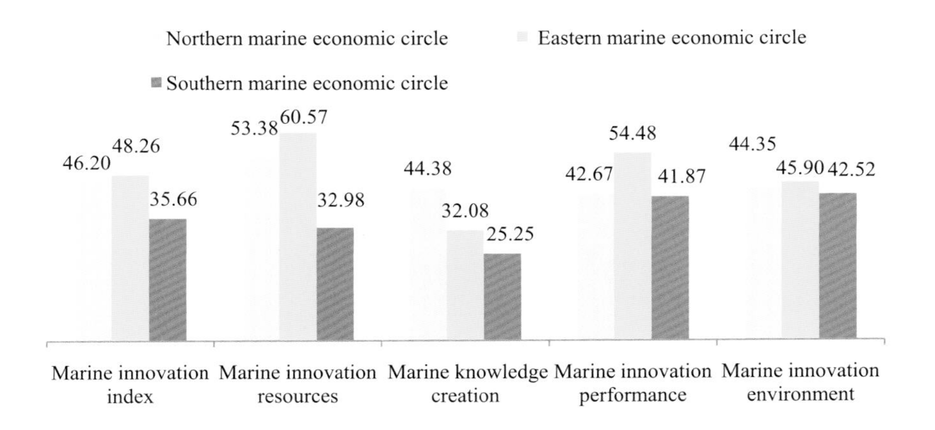

Figure 4-10 Scores of Marine Innovation Index and Sub-indexes of China's Three Marine Economic Circles in 2015

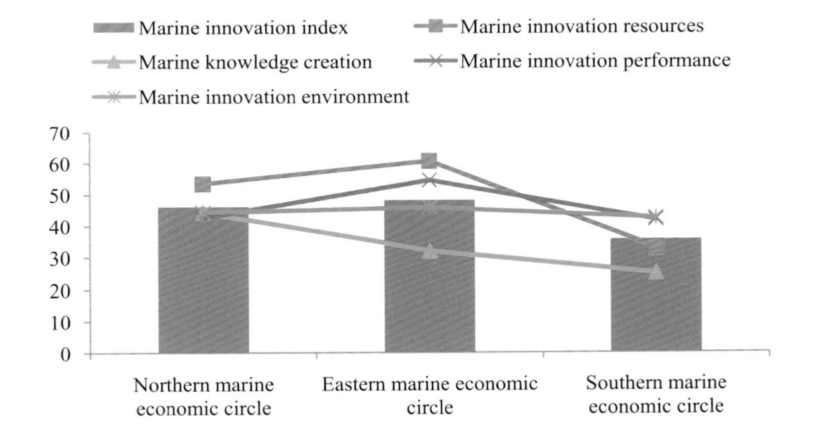

Figure 4-11 Relationship between the Regional Marine Innovation Index and Sub-indexes of China's Three Marine Economic Circles in 2015

The marine innovation index of the northern marine economic circle scored 46.20, which ranked the second among the three marine economic circles. Of the four sub-indexes, marine innovation resources which scored 53.38 made positive contributions to the marine innovation index. Marine innovation environment scored relatively low at 39.98. The low scores of the northern marine economic circle can be mainly attributed to the fact that the marine innovation environment is relatively weak and marine innovation development is yet to be improved. It's worth noting that the marine knowledge creation sub-index scored the highest among the three economic circles, thanks to the indicators like "S&T

works published in the current year" and "number of S&T papers published per 10 thousand scientific research personnel", indicating that the northern marine economic circle has a relatively strong marine S&T output and knowledge dissemination capability. In particular, there are a wide range of published S&T works.

Table 4-3 Regional Marine Innovation Index and Sub-indexes of China's Three Marine Economic Circles in 2014

Economic circle	Comprehensive index	Sub-index			
	Regional marine innovation index	Marine innovation resources	Marine knowledge creation	Marine innovation performance	Marine innovation environment
	A	b_1	b_2	b_3	b_4
Northern marine economic circle	46.20	53.38	44.38	42.67	44.35
Eastern marine economic circle	48.26	60.57	32.08	54.48	45.90
Southern marine economic circle	35.66	32.98	25.25	41.87	42.52

The marine innovation index of the southern marine economic circle scored 35.66, which was the lowest of the three economic circles. There were significant differences among the scores of the four sub-indexes. Marine innovation environment and performance achieved higher scores at 42.52 and 41.87. However, marine knowledge creation and marine innovation resources got lower scores at 25.25 and 32.98, which are the main factors leading to the lower marine innovation index. The lowest score of the southern marine economic circle means that there is considerable room for improvement. In the future process of marine innovation development, the innovation advantage of the Pearl River estuary and its two wings should be utilized to boost the development of the coastlines of Fujian, Beibu Gulf and Hainan Island by exploiting their regional advantage to achieve joint development and to radiate the economic development pattern driven by marine innovation to the entire southern marine economic circle.

V. Progress and Prospect of China's Marine Innovation Capability

President Xi Jinping emphasized that "we must develop marine science and technology and promote the transformation of marine science and technology towards a type of innovation leader. We should rely on S&T progress and innovation and strive to break through the technological bottleneck that hinders the development of marine economy and the protection of marine ecology. We should make an overall planning for marine S&T innovation". Innovation is the most important engine leading economic growth. Marine innovation is an important support which drives marine cause to make constant breakthroughs and realize steady and healthful development of marine economy.

Historically, the sub-indexes of marine innovation resources, marine knowledge creation, marine enterprise innovation, marine innovation performance and marine innovation environment have presented an upward trend. This trend has been fully borne out by the visible growth in the national marine innovation index. The average annual growth rate of 21.90% during the 2001-2015 period fully proves that the overall strength and competitiveness of marine technology have been constantly enhanced, independent innovation capability has been improved, marine innovation resources and knowledge output have been increasing substantially, marine innovation performance has become increasingly prominent and marine innovation environment has improved constantly as well.

National marine innovation capability is interconnected with marine economic development. The improvement of national marine innovation capability is correlated with the development of marine economy. During the period of 2012-2015, gross ocean product (GOP) and gross domestic product (GDP) showed a close growth rate and a big difference between them in previous years disappeared, indicating that national marine innovation capability has basically remained consistent with marine economic development and the contribution of marine innovation to economy has enhanced as well.

The indicators of the *"Outline of the National '12th Five-Year Plan' for the Marine Scientific and Technological Development"* are progressing well. In 2015, the proportion of GOP in GDP was 9.43%, the contribution rate of marine S&T developments and the transformation rate of marine S&T achievements reached 64.2% and 50.4% respectively, basically realizing the expected goals.

1. Interconnection between National Marine Innovation Capability and Marine Economic Development

National marine innovation capability is interconnected with marine economic development. Marine economy provides abundant funds for research and development of marine science and technology, thus improving the utilization efficiency of marine resources; the progress of marine science and technology and improvement of innovation capability, in turn promotes the growth of the marine economy and national economy. From 2001 to 2015, national marine innovation index, gross ocean product (GOP) and gross domestic product (GDP) all have shown a fluctuation trend (See Figure 5-1), with average annual growth rate of 21.90%, 14.82% and 14.07% respectively (See Table 5-1). The growth of national marine innovation index reached a peak in 2009 while GOP and GDP hit the lowest. The reason is that although the international financial crisis in 2008 brought huge negative impact on national economy and marine economy, China was pushing a huge number of enterprises to explore the ocean and expanding new space of marine economy through increasing investment, therefore marine innovation ushered in the development of great opportunities. When the impact of financial crises faded away and the macroeconomic situation became better, the national marine innovation index and its growth rate resumed a steady upward trend from 2009 to 2015, and both GOP and GDP have been on the rise. Meanwhile, during the period of 2012-2015, national marine innovation index, GOP and GDP have shown a close growth rate and the big difference among them in previous years has disappeared, indicating that national marine innovation capability has basically remained consistent with marine economic development.

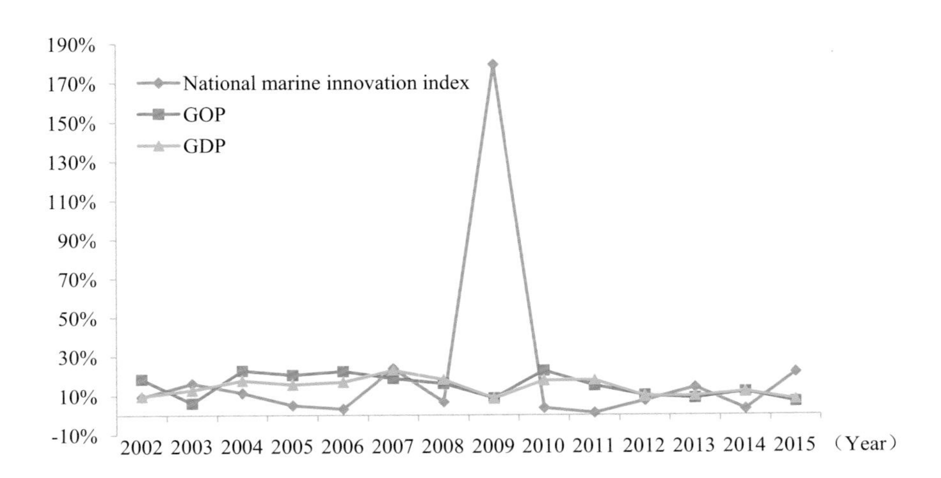

Figure 5-1 Growth Rate of National Marine Innovation Index, GOP and GDP from 2002 to 2015

Table 5-1 Growth Rate of National Marine Innovation Index, GOP and GDP (%)

Year	Growth Rate of the National Marine Innovation Index	Growth Rate of GOP	Growth Rate of GDP
2001	-	-	-
2002	9.45	18.41	9.74
2003	16.23	6.05	12.87
2004	11.41	22.67	17.71
2005	5.08	20.42	15.67
2006	3.25	22.30	16.97
2007	23.81	18.65	22.88
2008	6.77	16.00	18.15
2009	179.30	8.61	8.55
2010	3.70	22.60	17.78
2011	1.17	14.97	17.83
2012	7.40	10.00	9.69
2013	14.21	8.53	9.62
2014	3.07	11.76	11.83
2015	21.77	6.54	7.76
Average growth rate	21.90	14.82	14.07

2. Progress of the Indicators of the *"Outline of National '12th Five-Year Plan' for Marine Scientific and Technological Development"*

The *"Outline of National '12th Five-Year Plan' for Marine Scientific and Technological Development"*, the *"'12th Five-Year' Plan for National Marine Economic Development"* and other programs have set clear goals for marine innovation development during the *"12th Five-Year Plan"* period, aiming to guide China's marine innovation development of this period. At the end of the *"12th Five-year Plan"* period, carrying out data analysis about the completion status of those goals will be an important way to test the development of national marine innovation capability. Historical trend analysis of data and indicators obtained from the early stage of the *"12th Five-year Plan"* will facilitate the comprehensive review of the development in China's marine innovation, as is shown in Figure 5-1 and Table 5-2.

In 2015, the proportion of GOP in GDP reached 9.43%, very close to the expected goals and the contribution rate of marine S&T progress and the transformation rate of S&T innovation achievements reached 64.2% and 50.48%, smoothly realizing expected goals.

Table 5-2 Completion Status of Major Marine Indicators of the *"Outline of National '12th Five-Year Plan' Marine Scientific and Technological Development"*

Major indicators	The *"11th Five-Year"*	The goal of the *"12th Five-Year"*	Actual status	Completion status
Proportion of GOP in GDP		10%	9.43% (Year 2011—2015)	Close to Completion
Contribution rate of marine S&T progress	54.5%	>60%	64.2% (Year 2006—2015)	Completion
Transformation rate of marine S&T achievements		>50%	50.4% (Year 2000—2015)	Completion

Looking into the future, China needs to make greater effort to enhance the input in marine innovation resources, pay attention to the efficiency of marine innovation, make use of the supportive and leading role of marine innovation, transform the development mode of the marine economy, promote the upgrade of marine economic conversion and rely on marine science and technology to break through the constraints of energy, resources and environment during socioeconomic development, making marine innovation the core force for driving the development, transformation and upgrading of the marine economy so as to provide sufficient knowledge reserve and solid technological foundation for the construction of a maritime power.

VI. Specific Analysis on Regional Demonstration of China's Marine Economic Innovation Development

In order to respond to the requirements of "improving marine resources exploration ability, developing marine economy, protecting marine ecological environment, resolutely safeguarding national marine rights and interests and building maritime power" proposed in the report of the 18th CPC National Congress and *"Decision on Accelerating the Fostering and Development of Strategic Emerging Industries"* of the State Council, in May, 2012 the Ministry of Finance and State Oceanic Administration jointly issued notification which explicitly proposed that capital should be pooled to support Shandong, Zhejiang, Fujian and Guangdong as well as municipalities with independent planning status to carry out regional demonstration of marine economy innovation development through the development of strategic emerging industries such as marine biology. Then, in 2014, the Ministry of Finance and State Oceanic Administration approved the application of Jiangsu and Tianjin to become the demonstration areas of national marine economy innovation development and their implementation plans.

In 2015, the regional demonstration of our national marine economy innovation development made remarkable achievements and achieved remarkable results in achievements transformation and industrialization, public service platform for marine-related industries and special scientific research in marine-related public interest industries, thus enjoying a promising future.

1. Implementation Situation of Regional Demonstration in Shandong

Shandong is an ocean province with the coastline of 3345km, which accounts for 1/6 of the country's total aggregation and possesses 326 islands and more than 200 bays. The total sea area of Shandong amounts to 159,500 km², therefore, it enjoys unique advantageous natural environment for marine economy development. In recent years, under the strong support of the Ministry of Finance and State Oceanic Administration, Shandong has actively adjusted itself to the new normal of marine economic development, given prominence to local features, proactively innovated mechanism and support system, promoted the integration of production, university, research and application, and steadily pushed forward the regional demonstration work of marine economy innovation. It has made remarkable achievements, thus playing a leading role in promoting the development of economic development.

In 2015, the marine economy of Shandong achieved total output value of 12193 billion yuan, a year-on-year growth of 12.10%, accounting for 19.40% of the whole province's GDP. Marine biology and other strategic emerging industries achieved newly increased output value of 643 billion yuan, newly increased sales revenues of 457 billion yuan, newly increased taxes of 61 billion yuan, a year-on-year growth of 95.40%, 106.80% and 72.80% respectively.

In 2015, 96 regional demonstration projects of marine economy innovation development obtained the support of Shandong (including 63 industrialization projects, 33 platform projects), among which, 32 are new projects. It also cultivated 4 large-scale enterprise groups with annual output value exceeding 3 billion yuan, 7 enterprise groups with annual output value exceeding 700 million yuan, 23 medium-sized enterprises with annual output value exceeding 100 million yuan and 6 entrepreneurial type of enterprise centers with proprietary intellectual property rights. Three marine engineering equipment supporting products were developed with independent core intellectual property rights. Scientific research institutes and enterprises invested 8.5 billion yuan in R&D (including new projects), applied for (granted) 1133 patents, and transformed 234 technological achievements.

2. Implementation Situation of Regional Demonstration in Fujian

In 2015, the marine economy of Fujian achieved total output value of 6880 billion yuan, increasing by 10%. During the "*12th Five-Year*" period, the proportion of the total output value of marine economy in GDP of the whole province increased from 25% in 2010 to 27% in 2015.

The ecological environment has been significantly improved. Fujian takes the lead in implementing target accountability evaluation system to protect marine environment, setting indicators to control pollutants, carrying out monitoring and evaluation work to marine radioactive substances, and launching ten campaigns to protect marine ecological fishery resources. In 2015, the offshore sea areas with the second class water quality reached 66.10%, ranking first in the country.

The support capacity has been obviously strengthened. Since 2011, the whole province has issued 1201 certificates of the right to the use of sea areas and the total areas used have reached 33280 hectares, ensuring the smooth implementation of major marine projects. Fujian invested 1.4 billion yuan in constructing marine disaster prevention and mitigation projects known as "*Bai Qian Wan Project*" (Bai means hundred, Qian Thousand and Wan 10 thousand) , and established three-dimensional real-time monitoring network for the marine environment of the Taiwan Straits and the offshore areas of Fujian, the ability of marine observation and prediction having been remarkably improved. Emergency command system for the safety of fishery at sea has been completed and emergency management of fishing vessels and rescue capability have been further strengthened. Nighty four fishing ports of various kinds have been newly added, with the rate of fishing vessels taking shelter from the wind nearby increasing to 70%.

Innovation has increased remarkably. The strategy to develop marine economy through science and technology has been completely implemented, with leading products of fishery and the rate of main technology in place reaching up to 96%. The Island Research Center of the State Oceanic Administration, Southern Ocean Research Center of Xiamen and Virtual Oceanography Institute of Fujian stepped up the construction of marine science and technology innovation platform, cultivated a batch of marine S&T innovation enterprises. And the contribution rate of marine S&T progress reached 59.50% in 2015. It continued to deepen corporation and exchanges and took the lead in establishing China-ASEAN marine products trading platform. Xiamen was chosen as the site of China-ASEAN marine center.

Comprehensive management level has been improved significantly. Marketization of marine resources allocation has been further strengthened and there have been 82 cases of remising the right to the use of sea areas by bidding, auction and listing since 2012, with sea areas of about 4381 hectares. It is on the first batch of the country to make island protection plan, take market allocation mode to sell uninhabited islands openly to the public. It built marine law enforcement bases in Fuzhou, Dongshan and Pingtan and succeeded in completing the patrol missions assigned by the state to safeguard maritime rights. It carried out a joint enforcement initiative known as "*Blue Sword*" and the illegal incidents are decreasing year by year.

In 2015, Fujian province cultivated 56 marine enterprises above designated size, built more than 30 enterprise engineering technology centers of provincial and municipal levels, transformed 153 innovation achievements (with 19 being achievements from special funds for scientific research in marine public welfare industry, 16 being achievements from national 863 special funds, 42 being achievements from provincial special funds and 76 being achievements from independent R&D), applied for 143 invention patents and 80 utility patents, obtained 122 authorized patents, set 39 standards, 44 softwares, monographs, technical reports, etc, and published 15 academic papers.

3. Implementation Situation of Regional Demonstration in Guangdong

In 2015, the marine economy of Guangdong continued to develop steadily, achieving GOP of 1.52 trillion Yuan, an year-on-year increase of 10.50%, accounting for 20.90% of the whole province's GDP, maintaining its leading position nationwide for 21 consecutive years. The percentages of the first, second and third marine-related industries were adjusted from 2.4 : 47.5 : 5 in 2010 to 1.6 : 46.9 : 51.5 in 2015.

In 2015, in strict accordance to the demands of "*The Overall Work Plan of Guangdong Marine Economy Innovation Development Regional Demonstration*", Guangdong made great efforts to realize annual overall goals, ensuring that all targets including capital input, achievement transformation and industrialization, construction of the platform for industrial public services, renovation and restoration of sea areas and coastal zones, and the construction of collaborative innovation mechanism and industrial innovation ability were well fulfilled.

With regard to capital input, in 2015 Guangdong invested more than 6 billion Yuan in all fields of marine economy innovation development regional demonstration. Three new projects were added for implementing achievement transformation and industrialization and building platforms for industrial public services, with a project budget of 165 million Yuan.

With regard to renovation and restoration of sea areas and coastal zones, in 2015, Guangdong provincial government issued a plan on strengthening coastal protection and application. As a result, it obtained state fund of 273 million yuan to restore and renovate sea areas and islands and carried out a series of pilot projects on marine ecological restoration. According to the bulletin data about marine environment of Guangdong in 2015, the sea water quality of the offshore areas of the entire province was in good condition in general with the water quality of approximately 89.40% sea areas meeting the water quality standards of the first or the second class sea water, and the water quality of 3.70% sea areas being worse than the water quality of the fourth class sea water, which was 6.40% lower than that of 2014.

With regard to public service platform for marine industry, implementation of special projects for industrial public interests and construction of collaborative innovation centers, Guangdong was pushing forward in an all-around way to build a platform for Guangdong marine industrial public service in 2015, having completed 2067 marine natural product compounds with 1719 compounds added into the bank; having finished 1453 compound activity screenings, 568 biologic activity confirmations, 222 obvious activity confirmations; 11461 compounds information being recorded on information platform, among which 9608 having been reported publicly by literature. By the end of 2015, the actual funds of every project in place was 81.12 million yuan and the expenses of project funds were 65.67 million yuan and the funds expenses rate exceeded 80%. Guangdong created 24 new products (new materials / new techniques), published108 S&T papers, applied for 85 patents at home and abroad with 14 software copyrights. It was also awarded with two S&T prizes at and above provincial level, one second prize of State Technological Invention Award 2014, one excellence prize of China Patent Award 2013, and published one scholarly book.

With regard to industrial innovation capability, Guangdong have built 25 transformation bases of technological achievements in the fields of efficient and healthy cultivation of marine organisms, marine biological medicine and products and marine equipment and more than 55 industrial parks. It has also cultivated more than 20 enterprises with annual production value exceeding 1 billion yuan and built 3 clusters of marine strategic emerging industries. Collaborative innovation mechanisms and development model of integrating production with teaching and research are well established.

With regard to industrial development, in 2015, marine biology and other strategic emerging industries of Guangdong achieved newly increased production value of more than 100 billion yuan on the basis of 2012, with annual sales revenues of the industry reaching 267.1 billion yuan and annual taxes exceeding 6.3 billion yuan. Both the annual growth rate of marine strategic emerging industries like marine biology and the total amount of marine economy maintained the leading position in the country. The structure of marine industry and space layout were further optimized and the percentages of the first, second and third marine-related industries were adjusted to be 1.6 : 46.9 : 51.5 in 2015.

4. Implementation Situation of Regional Demonstration in Jiangsu

In 2015, the demonstration region of Jiangsu marine economy innovation development made an annual total investment of 858 million yuan (including 21 fiscal supporting projects), among which, the R&D input was 69 million Yuan, achieved newly increased output value of 3.01 billion yuan, sales revenues of 2.138 billion yuan, profit and tax of 262 million yuan, applied for 262 patents, transformed more than 40 achievements, created 117 new products (new techniques/ new equipment/ new technologies/new methods), set 76 standards(software/monograph/technical report), cultivated 272 talents with senior professional title, built 39 enterprise R&D centers above municipal level and engineering centers, won 13 S&T prizes of national and provincial levels, and cultivated 13 enterprises above designated size.The output value of marine equipment and desalination industry of the whole province reached 45 billion Yuan.

With regard to achievements transformation and industrialization, in 2015, Jiangsu approved 19 achievements transformation projects (7 projects were supported last year), including 3 sea water desalinations, 12 marine equipment and facilities, with planned investment totally 810 million yuan. By the end of March, 2016, Jiangsu has made annual total investment of 823 million yuan, achieved sales revenues of 2.088 billion yuan and profit and tax of 251 million yuan, overfulfilling target tasks.

With regard to marine public service platforms, in 2015, Jiangsu allocated fiscal funds of 28 million yuan to support the construction of "Jiangsu public service platform for marine equipment innovation" led by No.702 Research Institute of China Shipbuilding Industry Corporation and "Jiangsu digital design and manufacturing technology service platform of marine equipment" led by Jiangsu University of

Science and Technology, with two years as implementation period. In 2015, focusing on making marine equipment industry bigger and stronger, the two platforms carried out a series of work in advancing the integration of production, teaching and research, breaking through technological bottleneck, and cultivating high-quality talents. It made an annual total investment of 34.59 million yuan, carried out activities such as training, consulting, technology transfer, design development, testing and experiment of more than 300 times, provided services to nearly 200 enterprises and public institutions of marine engineering equipment inside and outside Jiangsu province, cultivated 105 senior technical staff, applied for 74 patents, created 35 new products/new techniques/new equipment/new methods, etc, set 39 standards (software/monograph/technical report), won one second prize of National S&T Progress Award, one second prize of National Invention Award, and four prizes of Provincial S&T Award.

5. Implementation Situation of Regional Demonstration in Tianjin

In 2015, the total output value of Tianjin marine economy was 550.6 billion Yuan, an increase of 9.50%, approximately accounting for 33.30% of the city's total output value. Marine economy scale per unit of coastline output reached 3.5 billion Yuan, being at the forefront of the country. Marine strategic emerging industry achieved an added value of 63 billion yuan, which accounted for 11.50% of the city's total output value with average annual growth rate of 19%, becoming the important growth point of the city's marine economy. Tianjin took the lead in the technology and capability of seawater utilization, and the scale of installation capacity of seawater desalination reached 317 thousand tons per day, approximately accounting for one third of the country. Marine salt industry and marine chemical industry have been transformed and updated constantly driven by seawater utilization industry. In 2015, the seawater comprehensive utilization industry and related industries realized added value of 480 million yuan, and marine equipment industry achieved an added value of 56.5 billion yuan.

In 2015, the regional demonstration of Tianjin marine economy innovation development started and implemented 43 key projects approved by the state and planned to invest 935 million Yuan. The actual completed investment was 569 million Yuan with an investment completion rate reaching 60.80%. Tianjin also completed transforming 46 achievements, applied for 303 patents, added 33 new technologies and techniques and 46 new industrial technical standards. Based on projects, it built 42 R&D centers, engineering centers and pilot bases, of which, three R&D centers are of national level, twelve are of municipal level and ten are enterprises ones. One engineering center is of national level, nine are of municipal level and two are enterprise ones, and five are industry pilot bases. All indicators were above the same period last year and exceeded the implementation requirement to the projects.

In 2015, the added value of seawater comprehensive utilization industry in Tianjin was 480 million yuan, of which 160 million yuan was from seawater desalination and direct use of seawater, 320 million yuan was from salt manufacturing from concentrated seawater and chemical engineering and other comprehensive utilization. Marine equipment manufacturing industry achieved an added value of 56.5

billion yuan. In 2015, Tianjin approved 11 projects on marine development driven by S&T, obtained special funds from financial support of 6.8 million yuan with matching funds from enterprises and scientific research institutes of 25 million yuan, which is expected to create economic benefits of 300 to 400 million yuan.

6. Implementation Situation of Regional Demonstration in Zhejiang

In 2015, the GOP of Zhejiang was 618.4 billion Yuan, up by 7.33% over the previous year, 0.5% higher than the current price growth rate of national economy, increasing by 63.20% compared to 2010, with average annual growth rate of 10.30%, of which, the first industry was 44.783 billion Yuan, the secondary industry 234.273 billion Yuan and the tertiary industry 338.986 billion Yuan, up by 4.74%, 3.54% and 10.49% respectively over the previous year. The proportion of marine economy in the total output value was 14.41% with a slight increase. The percentage of the first, second and tertiary industry structure was 7.2:37.9:54.9.

In 2015, Zhejiang focused on the construction of the ten leading industrial agglomeration areas of provincial level along the coast. It planned and established ten industrial agglomeration areas at provincial level along the coast and became the major platform for the development of marine industry along the coast. In 2015, the ten leading industrial agglomeration areas at provincial level along the coast realized total industrial output value of 543.22 billion yuan, up by 40% over last year, achieving service revenues of 254.60 billion yuan, increasing by 54.80% compared to last year. By the end of 2015, the ten industrial agglomeration areas at provincial level along the coast have totally introduced 301 industrial projects with every investment exceeding 100 million yuan.

In 2015, Zhejiang provincial government issued "*Zhejiang Major Project Construction Plan of Marine Economic Development 2015*" and approved 479 projects with total investment of 868 billion yuan, of which, 124.7 billion yuan was planned to be invested in 2015. By promoting the construction of major projects for marine economy development to stimulate continuous increase of investment in projects of marine economy, it is estimated that the seven coastal cities will have made effective investment in marine economy of more than 230 billion yuan.

VII. Specific Analysis on the Input-Output Efficiency of Marine Science and Technology in China

Globally, marine science and technology have become one of the most critical areas among maritime powers competing for maritime leading position and their international discourse rights. Though researches on marine science and technology in China started relatively late, great-leap-forward developments have been made after several decades' exploration and growth. The development of marine science and technology has entered a stage of qualitative breakthrough from quantity accumulation of early periods. Under the new normal, rationally allocating marine science and technology resources and increasing its input-output efficiency are powerful driving forces in pushing forward the development of the marine science and technology in China.

Innovation is the first driving force guiding developments. The vision of innovative, coordinated, green, open and inclusive development was put forward at the Fifth Plenary Session of the 18th Central Committee of the CPC, with "innovative vision" as the top priority. The "*13th Five-Year*" is both a strategic opportunity in realizing the leap-forward development of marine science and technology and a key stage in promoting marine S&T innovations into a new height. To analyze and forecast at national level based on the review and summary of the characteristics of input-output efficiency of marine science and technology since the "*10th Five year*" will be of both practical and realistic significance to coordinately push forward the growth of marine science and technology of the "*13th Five-Year*".

In the light of the above discussion and with the city as basic research unit, marine science and technology input-output efficiency as research subject, data of the marine scientific research institutes during the period of 2001-2015 as the research base, and the DEA as the model, this chapter has done calculation and measurement to the input-output efficiency of marine-related cities and then explored the rules of it. On this basis, the chapter will do retrospective analysis and trend prediction to the input-output efficiency of marine science and technology from the "*10th Five-Year*" period to the "*13th Five-Year*" period in China, in order to provide a data support in promoting the development of marine science and technology.

1. Summaries and Tendency Forecasts

Marine science and technology are a significant engine in pushing forward the implementation of China's marine strategies and the development of the blue economy. In spite of the advances of the maritime power strategies and continuous strengthening of marine R&D, the existing national uneven resource allocation has resulted in significant differences among marine science and technology input-output efficiency. Based on the data of input-output efficiency of marine science and technology and the software of DEAP2.1, the overall input-output efficiency of marine science and technology, the pure technical efficiency and scale efficiency have been calculated and measured (See Appendix 8 for Calculating Methods & Processes) with features and rules of results as follows:

1.1 Space scope features the layout of "high in the north and east while low in the south".

According to the geographic distribution, the marine-related cities in China can be divided into five regions, which are the northern, eastern, southern, southwestern and middle western ones, among which the northern regions include Beijing, Tianjin, Shandong Province, Hebei Province, Liaoning Province and Heilongjiang Province; Shanghai, Jiangsu Province, Zhejiang Province and Fujian Province belong to the eastern regions; the southern regions consists of Guangdong Province and Hainan Province while the southwestern coastal regions refers to the Guangxi Zhuang Autonomous Region. Some inland regions do possess rich marine scientific research strength, but their locations and data show great differences when comparing with the Guangxi Zhuang Autonomous Region. As a result, they are listed into the middle west coastal regions. The existing data available from 2001 to 2015 involves totally 59 marine-related cities. Based on the rules mentioned above, there are 23 marine-related cities in the north, 17 in the east, 13 in the south, and 3 in southwest and middle west respectively (See Appendix 9 for Classification principle and the list of marine-related cities.

In general (See Table 7-1), marine science and technology input-output efficiencies in the north, east, south and the southwest marine-related regions are relatively low and fail to reach the ideal state with great disparities among the four regions. Vertically, the efficiency in the northern region has been higher than others, and the efficiency in the eastern region has been higher than those in the southern and southwestern ones, while the overall efficiency in the south has been kept at a low level due to the rapid increase in marine S&T input in the context of insignificant increase of output, resulting in low input-output efficiency of marine S&T input. The efficiency in the southwestern coastal area is generally higher than that in the south. On average, the overall efficiencies in different regions from 2006 to 2015 are 0.477, 0.426, 0.257 and 0.358, respectively. It can be found that the north enjoys great superiorities, followed by the east and the southwest, while great gaps exist between the south and the other three regions.

Though Wuhan, Lanzhou and Xi'an are inland cities, their marine science and technology input-output efficiencies have been operating at a high level, among which the efficiencies from 2006 to 2009 are the highest and become stable with the average efficiency at above 0.94 despite a slight decrease after 2010. The marine science and technology input-output efficiency of these cities is distinctly competitive compared with other four regions.

Thus it can be seen that there exist evident spatial gradient disparities in the input-output efficiency of China's marine science and technology. The efficiency in the north is the highest, the east ranks the 2nd, while the efficiencies in the southern and southwestern regions are relatively low with the southwestern ones a bit higher than that in the southern regions. And the efficiency in the middle west coastal areas has been operating at a high level for a long term. As a consequence, China's marine science and technology input-output efficiency is characterized by a geographical layout of "high in the north and the east while low in the south" .

Table 7-1 The Overall Input-output Efficiency of Marine Science and Technology

Region /Year	North	East	South	Southwest	Middle west
2006	0.454	0.397	0.176	0.474	1
2007	0.463	0.419	0.206	0.459	1
2008	0.465	0.415	0.23	0.407	1
2009	0.497	0.425	0.269	0.355	1
2010	0.484	0.396	0.268	0.293	0.976
2011	0.505	0.421	0.272	0.291	0.965
2012	0.487	0.429	0.26	0.292	0.954
2013	0.472	0.451	0.269	0.335	0.949
2014	0.472	0.445	0.297	0.335	0.943
2015	0.468	0.46	0.327	0.343	0.977
average	0.477	0.426	0.257	0.358	0.976

Note: Raw data have been smoothed with five-displacement moving average method

1.2 Disparities are decreased among regions in marine science and technology input-output efficiency in time scale.

Geographically, the overall efficiencies decrease regionally in a time scale (See Figure 7-1). From 2006 to 2015, the marine science and technology input-output efficiency of the northern regions has been higher than other ones, keeping a long-term stability of above 0.45; the east has been ranking second and remaining above 0.40 after 2010, which shares little difference compared with the north. The marine science and technology input-output efficiencies of the south and the southwest are relatively low, while the disparities between them are getting closer. After 2010, these two regions share little difference with each other, with disparities from the northern and the eastern regions being decreased gradually. It can be found in Figure 7-1 that the floating range of the four regions has been decreased to 0.30-0.50 from 0.15-0.50 in 2006, especially that the disparities between the north and the east and between the south and the southwest coastal areas are distinctly reduced.

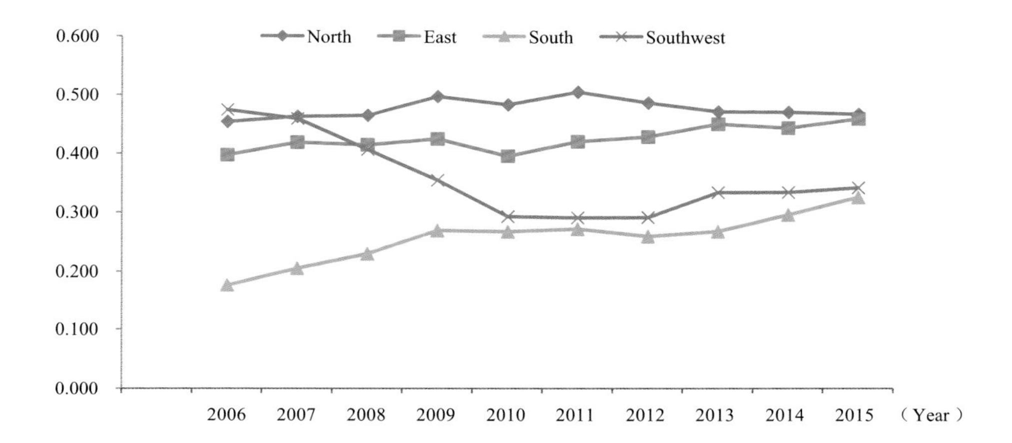

Figure 7-1 Overall Input-output Efficiency of Regional Marine Science and Technology from 2006 to 2015

1.3 The advantages of administrative locations benefit the gathering of marine S&T resources, while input redundancy exists in some cities.

The major marine-related cities in China include municipalities directly under the central government, municipalities with independent planning status and provincial capitals, which enjoy advantages in administrative locations. As can be seen from Table 7-2, the advantages of administrative locations play a crucial role in allocating marine S&T resources. From 2001 to 2015, the proportion of marine S&T input in funds and personnel of major marine-related cities in the whole country have been changed to 93.84% and 89.33% from 94.80% and 90.27% respectively. Compared to 2001, though the proportion of marine S&T input decreased, resources input still had absolute advantages. That means under the dominance of the administrative locations, the marine science and technology resources in China are mainly concentrating on municipalities directly under the central government, municipalities with independent planning status and provincial capitals which are listed in marine-related cities.

Table 7-2 Proportions of the Marine S&T Resources Input of Marine-related Cities in 2001 and 2015

City	Capital input in marine science and technology		Personnel input in marine science and technology	
	2001 Proportion(%)	2015 Proportion(%)	2001 Proportion(%)	2015 Proportion(%)
Municipalities directly under the central government	43.10	53.07	40.48	47.06
Municipalities with independent planning status	26.96	15.02	24.75	11.18
Provincial capitals	24.75	25.74	25.04	31.09
Other marine-related cities	5.19	6.16	9.73	10.67

By calculating the marine science and technology efficiency of the marine-related cities from 2001 to 2015, it can be found that the advantages of the administrative locations will not only accelerate the gathering of the marine S&T resources, but also will significantly influence the marine S&T input-output efficiency, which are manifestated as follows:

(a) The traditionally powerful marine-related cities enjoy high marine S&T developments, while under the dominance of the administrative superiority, apparent redundancy exists in Beijing, Shanghai, Guangzhou and Tianjin. The Table 7-3 shows the DEA calculating results of China's eight traditional powerful marine-related cities. In the respect of the scale efficiency, Beijing, Shanghai, Guangzhou and Tianjin were significantly lower than that of other four cities. In most of the years they were in a stage of decreasing returns to scale, especially in Beijing, Shanghai and Guangzhou, of which the scale efficiencies were 80% below the optimal level, which resulted in severe redundancy. The pure technical efficiencies of Beijing and Guangzhou reached 80% above the DEA efficiency, however, the scale efficiencies were not relatively low. So conclusion can be drawn that the mismatch of the science and technology input and output is the major reason for unsatisfactory overall efficiency. On the other hand, the scale efficiency of Tianjin was relatively effective, but the pure technical efficiency was only 0.523, indicating that there exists problems in current marine S&T resources management and levels. By contrast, the pure technical efficiency and scale efficiency of Shanghai are both between 70% and 80% of the optimal level. Although there existed some problems in input scale and management, they were not serious. From here we can see that the advantages in administrative locations do bring abundant benefits in marine S&T technology resources input to these cities, but at the same time, they lower the input-output efficiency of marine science and technology.

(b) Qingdao, Dalian and Xiamen are municipalities with independent planning status and Hangzhou is a provincial capital. All of them have the advantages in administrative locations and at the same time, all of them are traditionally powerful marine-related cities. What's more, all these four cities reached above 80% of the optimal level in terms of the scale efficiency, particularly Dalian, Hangzhou and Xiamen whose scale efficiency reached 90% of the optimal level and marine S&T resources input got extremely close to the optimal ones. However, the returns to scale of Qingdao, Hangzhou and Xiamen have been decreasing for a long term, while returns to scale of Dalian has been increasing for a long period, which means the marine S&T resource input could be further optimized. With regard to the pure technical efficiency, Qingdao enjoyed the best level, while the pure technical efficiency of Dalian, Hangzhou and Xiamen were all less than 60% of the DEA efficiency. Therefore, a significant way to realize the marine S&T input-output transformation in the three cities mentioned above is to promote S&T management and levels. Given the above, advantages like superior initial development condition and early start of marine S&T innovation make the scale efficiencies of these cities close to the optimal level. Generally, compared to cities with high marine S&T input-output efficiency like Qingdao, measures need to be taken by Xiamen, Hangzhou and Dalian to increase their input-output efficiencies.

(c) The marine S&T input-output efficiencies of the provincial capitals like Nanjing, Wuhan, Lanzhou and Xi'an are in a relatively effective level. From 2001 to 2015, the DEA efficiencies of Nanjing, Wuhan, Shijiazhuang, Lanzhou and Xi'an had exceeded 80% of the optimal level, among which Lanzhou and Xi'an had always been in the optimal level. Moreover, the returns to scale of the five cities remained unchanged for more than half of the time, thus effectively realizing the transformation of marine S&T input into output. Accordingly, conclusion can be made that the advantages of administrative locations bring absolute policy supports to these five cities, which not only drastically has pushed forward the growth of the marine science and technology, but also realized the effective marine S&T input-output transformation.

Table 7-3 Analyses on the DEA Calculating Results of Marine S&T Input-output in Traditionally Powerful Marine-related Cities

City	Average value of overall efficiency	Average value of pure technical efficiency	Average value of scale efficiency	Number of times of crs	Number of times of irs	Number of times of drs
Beijing	0.659	0.906	0.722	2	0	13
Shanghai	0.542	0.735	0.766	1	0	14
Guangzhou	0.629	0.889	0.704	2	0	13
Tianjin	0.431	0.523	0.817	0	1	14
Qingdao	0.835	0.999	0.836	4	1	10
Dalian	0.353	0.405	0.910	2	10	3
Hangzhou	0.505	0.565	0.906	1	5	9
Xiamen	0.462	0.494	0.934	2	6	7

1.4 Spatial differences exist in marine S&T resources allocation.

The gradient disparities in marine S&T input-output efficiency among marine-related cities lie in the disparities between marine S&T management level and the resource input scales. Based on the calculating results and according to the effective degree of the efficiency of the pure technical efficiency and scale efficiency, marine-related cities in China are classified into categories as follows (See Table 7-4):

Table 7-4 Classifications of Marine-related Cities in China Based on the Differences in Pure Technical Efficiency and Scale Efficiency

Type	Basis of classification	Typical city
High-efficiency city	Scale efficiency >0.8, Technical efficiency >0.8	Qingdao、Nanjing、Shijiazhuang、Wuhan、Nanning、Lanzhou、Xi'an
Technical-efficiency city	Scale efficiency >0.8, Technical efficiency >0.8	Beijing、Shanghai、Guangzhou、Shenzhen、Dandong、Shaoxing、Putian、Quanzhou、Ningde、Weifang、Zhuhai、Shantou、Jiangmen、Jinan、Beihai、Foshan、Weihai、Haikou、Dongying、Jieyang、Rizhao
Scale-efficiency city	Scale efficiency >0.8, Technical efficiency >0.8	Tianjin、Dalian、Xiamen、Shenyang、Zhoushan、Hangzhou、Nantong、Wenzhou、Yantai、Yancheng、Nantong
Low-efficiency city	Scale efficiency >0.8, Technical efficiency >0.8	Jinzhou、Yantai、Zhanjiang、Ningbo、Fuzhou

(a) The type one cities in the figure refer to the marine-related cities that are relatively effective in marine S&T management level and resources input scale, namely, both the pure technical efficiency and scale efficiency are above 80% of the DEA efficiency. The major cities include traditional marine-related cities like Qingdao and Nanjing as well as the provincial capitals like Wuhan and Lanzhou. Advantages in administrative locations, policy supports and solid developing foundations make the marine S&T input-output efficiency of these cities get closer to the optimal levels and realize effective input-output transformation. The scale economies effects of these cities have all been maintaining unchanged or close to stability in returns to scale, which reflects that the marine S&T resource input has been lacking of elasticity. Attention needs to be paid to the connotative development of marine science and technology utilization while appropriately adjusting marine S&T resource input in this area. In a word, efforts need to be made in intensifying the utilization efficiency of the science and technology resources to promote

the output with the existing resource input.

(b) The type two cities refer to the marine-related cities with relative effectiveness in marine S&T management level, whereas the resource input scale is not that ideal. In short, the pure technical efficiency of the type two cities is above 80% of DEA efficiency, while the scale efficiency is below 80% of the optimal level. Beijing, Shanghai, Guangzhou and Shenzhen are included in these cities and what's more, the marine S&T management level of these type two cities are high but certain disparities exist between resource input scale and optimal level. What makes it different is that with the advantages in administrative locations, Beijing, Shanghai and Guangzhou have excessive input in marine science and technology and become marine-related cities that possess effective technology but input redundancy. The emerging marine-related cities like Shenzhen, Zhuhai and Jinan, with the input-output of marine science and technology mainly relying on incremental increase of returns to scale, belong to the marine-related cities with effective technology but less input. In addition, it can be found that the marine-related cities in pan-pearl-river delta region are mainly technical efficiency cities. As a result, conclusion can be made that it is a vital way to continuously increase marine S&T resources input in promoting input-output efficiency of southern areas of China.

(c) The type three cities refer to the marine-related cities in which the marine S&T resource input is in a relative effectiveness status but there exists differences between science and technology management level and optimal level, namely, the scale efficiency of these cities is above 80% of the DEA efficiency, while the pure technical efficiency is below 80% of the optimal level. The major cities of this type are Tianjin, Xiamen, Zhoushan, Hangzhou and Nantong. What is noteworthy is that the scale-efficiency cities mainly refer to marine-related cities in Yangtze River delta area, and the scale economies effect of these cities had been close to constant returns of scale. To improve the resource utilization level of the current input scale shall be regarded as the top priority in increasing the marine S&T input and output efficiency in the Yangtze River delta area in the years ahead.

(d) The type four cities refer to the marine-related cities whose marine S&T management level and resource input scale have great disparities from the optimal level, which means that both the pure technical efficiency and the scale efficiency are 80% less than the optimal level. The type four cities mainly consist of Zhanjiang and Ningbo, which are located in China's circum-Bohai-Sea, Yangtze River Delta and Pearl River Delta. Efforts need to be made in the full utilization of the geographic advantage and the dramatic development to push forward the marine S&T input-output efficiency.

1.5 The "*13th Five-Year* " possesses broad prospects.

The "*13th Five-Year*" is a key period in realizing marine S&T strategic breakthroughs, while marine economy urges more marine innovations during its development. To conduct retrospective analysis to marine S&T input-output efficiency from the "*10th Five-Year*" to the "*12th Five-Year*" and grasp the overall developing trend are of absolute significance in coordinately promoting marine S&T development in the "*13th Five-Year*" period.

With the state as the decision-making unit of DEA evaluation, we have calculated national marine S&T input-output overall efficiency from 2001 to 2015 and done predictions about it by means of trend analysis approach. The results are shown in Figure 7-2. Seen in an all-round way, the marine S&T input-output efficiency in China has been rising with fluctuations. From the view of the average moving results, the overall efficiency has experienced a trend of significant increasing-slight decreasing-stable recovery. Before 2008, the marine S&T input-output efficiency presented dramatic rise trend with the overall efficiency increasing by 90% relatively. The research group owes this to China's entry to the WTO in 2001 and as a result, a new opening up pattern that is all-dimensional, multi-level and wide-ranging came into being. The "strategy of implementing marine exploration" put forward by the 16th National Congress of the Communist Party of China brought great opportunities for the development of the marine industry and the rise of marine science and technology, which led to the fast growth in marine S&T input-output efficiency. The overall efficiency dropped by around 17% during the period from 2008 to 2012. Based on the data mentioned, the main reason is the hysteresis of the increase of the marine S&T input-output efficiency, which would give rise to continuous low efficiency of the marine science and technology for a short term. After 2012, the overall efficiency enjoyed a rising trend and the marine S&T input-output efficiency in 2014 and 2015 reached the DEA efficiency. As the 18th National Congress of the Communist Party of China released the strategy of becoming a maritime power, the marine science and technology has been put onto an unprecedented height, and the marine S&T input-output efficiency has also been improved further. According to the tendency curves that reflect the average movements, the overall efficiency from 2016 to 2020 will be further improved, and marine S&T development will gradually enter into a new strategic stage that realizes qualitative breakthroughs.

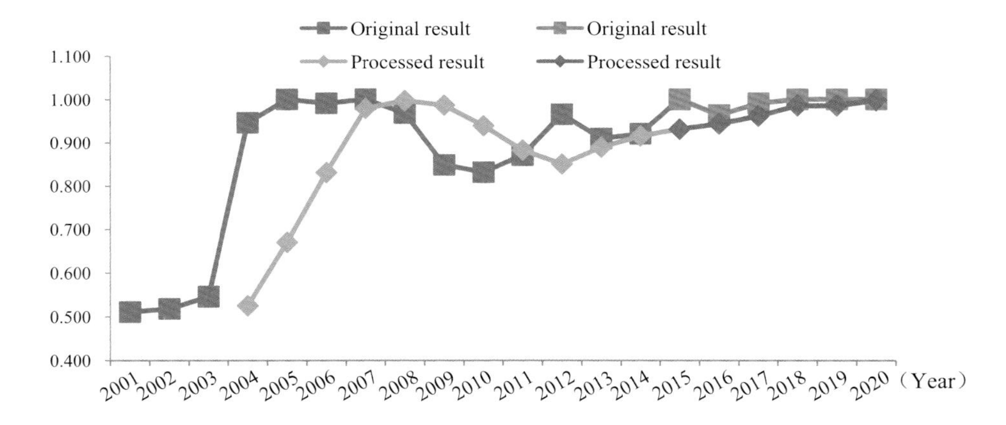

Figure 7-2 Overall Efficiency of National Marine S&T Input-output from 2001 to 2020

The average values of marine S&T input-output overall efficiency during the "*10th Five-Year*" and the "*13th Five-Year*" have been calculated respectively based on the analysis of the overall efficiency from 2001 to 2020 (See Table 7-5). The result shows that except the "*10th Five-Year*", the overall efficiencies of the other three periods have all reached above 90% of the optimal level. Therefore, it can be predicted that the average value of the overall efficiency during the period of the "*13th Five-Year*" will reach 0.991. In terms of the growth range, the overall efficiency increased by 0.54% and 6.22% respectively during the periods from the "*11th Five-Year*" to the "*12th Five-Year*" and from the "*12th Five-Year*" to the "*13th Five-Year*", which reflects from one side that prosperous progress has been made in promoting marine S&T input-output efficiency since the top priority had been put onto the marine industry in the 18th National Congress of the Communist Party of China. The promising prosperity of the marine science and technology in the "*13th Five-Year*" is foreseeable from Figure 7-2 and Table 7-5.

Table 7-5 The Average Value of the Overall Efficiency of Marine S&T Input-output from the "*10th Five-Year*" to the "*13th Five-Year*"

Period	The 10th Five-Year	The 11th Five-Year	The 12th Five-Year	The 13th Five-Year
1	0.511	0.991	0.870	0.964
2	0.518	1.000	0.965	0.991
3	0.546	0.968	0.909	1.000
4	0.946	0.848	0.920	1.000
5	1.000	0.831	1.000	1.000
Average	**0.704**	**0.928**	**0.933**	**0.991**

2. Countermeasure and Suggestion

The marine science and technology in China now are in the most satisfactory window period in realizing great-leap-forward developments with more opportunities to be grasped than the challenges to be dealt with. However, the great disparities in national marine S&T resource allocations and input-output efficiencies among cities and regions resulting from administrative location advantage and policy support will still exist for a long time in the future. As a result, how to realize effective transformation of marine S&T input-output has been a key issue in pushing forward marine science and technology to achieve an innovation-oriented transformation.

(a) Based on the "*National Science and Technology-oriented marine Development Plan (2016-*

2020)" released on December 12th, 2016, and guided by the policies of the "*Outline of the National Innovation-Driven Development Strategy*" and the "*National '13th Five-Year' Planning for Scientific and Technological Innovation*", we should deeply understand the new requirements of major strategic deployment for marine science and technology. We need to improve the top-level design of the marine innovation strategy and protect the policies and systems of the marine science and technology innovation at a national level so as to realize the great-leap-forward developments of the marine science and technology input-output efficiency. Considerations need to be taken into account in giving the southern and the southwestern coastal areas appropriate policy supports and accelerating the flow and redistribution of the elements such as capital, talents, technology and information, based on the current status of marine S&T input-output efficiency that features the "high in the north and east while low in the south" layout.

(b) The spatial disparities of marine S&T resource allocation among regions and the gradient disparities of marine S&T input among cities should be narrowed down to encourage the transfer of advanced marine science and technology in the developed regions to the less developed regions. We should allocate regional marine S&T force properly and construct the marine science and technology sharing platform like marine virtual institutes. Generally speaking, the marine science and technology of the traditionally powerful coastal city are well-developed as a whole. Efforts need to be made in improving effectiveness and optimizing the utilization environment of science and technology while appropriately adjusting the scale of the marine S&T input. Besides, most of the emerging marine-related cities are non-high-efficiency cities, whose marine S&T input-output efficiency are operating at a low level. Therefore, measures need to be taken such as giving the full play to the radiation and driving function of the city circle in circum-Bohai-Sea, Yangtze River Delta and Pearl River Delta, thus promoting the utilization of the marine S&T resources by talents flow and technology transfer.

(c) Coordinated cooperation in policy-industry-university-research need to be strengthened among marine-related cities to gradually change the situation that the marine S&T research and development are done by the research institutes or the universities individually. Enterprises, institutes as well as the universities should be encouraged to jointly build the science and technology R&D centers, and share marine S&T patents by means of capital and technology investment. As for the policy making, local governments should introduce related policies to stimulate the enthusiasm of S&T personnel, improve the protection system for the intellectual property, and accelerate the construction of the achievement transformation platforms so as to realize the seamless connection of the marine S&T resource input, achievement transformation and marketization, promote the virtuous cycle of the marine S&T input and output as well as to increase the rate of marine S&T input-output transformation and marketization of the achievements and to give full play to the economic and social effectiveness of the marine S&T resources.

3. Conclusion

Marine S&T strength is a supporting force in pushing forward the development of marine economy, and the marine S&T input-output efficiency plays a decisive role in realizing the marine economic and social development goals. Based on DEA model, the research has done calculations and analysis on the marine S&T input-output efficiency of the national marine-related cities during the period from 2001 to 2015. Meanwhile, the research has done retrospective analysis of the data from the "*10th Five-Year*" to the "*12th Five-Year*" and made predictions for the "*13th Five-Year*" on that basis. What needs to be mentioned is that to analyze the features and tendency of marine S&T input-output efficiency is a long time process, which calls for the support of the panel data of several continuous time sections to accurately reflect the rules of the resource utilization efficiency. As a result, a long-term monitoring of the marine S&T input-output efficiency of marine-related cities in national wide and a series of optimizing ways to be put forward will be the research directions in the days to come.

Appendix

Appendix 1: Indicator System of National Marine Innovation Index

1. The Connotation of National Marine Innovation Index

National marine innovation index is a comprehensive index, which measures marine innovation capability and fairly reflects the marine innovation quality and efficiency of a country.

By using both Chinese and international theories and methodologies for evaluating a country's competitiveness and innovation, and based on the connotation analysis of innovative marine powers, we have identified the indicator selection principles and constructed an indicator system of national marine innovation index in five aspects, which are marine innovation resources, marine knowledge creation, marine enterprise innovation, marine innovation performance, and marine innovation environment, aiming at reflecting the characteristics of China's marine innovation capability at different levels comprehensively, objectively and accurately, and forming a full-fledged indicator system and evaluation methods. The index measurement provides technical support and consulting service for comprehensively evaluating the progress of building an innovative marine power and perfecting marine S&T innovation policies.

2. The Connotation of Innovative Maritime Powers

In order to be a maritime power, China needs to promote the transformation of marine science and technology for the purpose of leading innovation. International historical experiences show that the development of marine science and technology is the fundamental guarantee for a maritime power. Therefore, we should establish comprehensive evaluation indicator system for national marine innovation, examine marine development trends from a strategic perspective, strengthen basic marine research and the construction of talent teams, vigorously develop marine science and technology to provide decision support for various aspects of economic and social development.

The evaluation of national marine innovation index will enable national and local governments to get informed of the progress situations in the implementation of marine S&T development strategies and identify potential problems in timely fashion so as to provide basic information for further decisions. It will allow national and international public to understand the marine cause of China in terms of the progress and accomplishments that it has achieved, the trends and the problems that it may encounter. It will also allow enterprises and investors to study and assess opportunities and risks in marine field and provide relevant information for scholars and institutes engaged in marine research.

China has undergone three stages throughout the history of marine economic development: resources dependence, extensive expansion of industrial scale and a stage of quantity-to-quality shift. Rapid development of marine science and technology promotes the constant expansion of marine industry and, in itself, becomes a new growth point of marine economy. China has vast sea areas and

abundant marine resources but years of extensive development have made resources and environment problems increasingly prominent, hindering further development of marine economy. Thus, only constant marine innovation can help promote the healthy development of a marine economy and help turn China into one of the world's innovative marine economic powers.

The most distinctive feature of "innovative maritime powers" is that there is a fundamental change in the development model of national marine economy compared to traditional development model. To tell whether a country is an innovative maritime power or not, we should analyze whether its marine economic growth is mainly driven by factors (the traditional model of natural resources consumption and capital utilization) or is based on innovation activities featuring creation, dissemination and application of knowledge.

Innovative marine powers should have the following four capabilities:

(a) Higher input capability in marine innovation resources;

(b) Higher marine knowledge creation, dissemination and application capability;

(c) Better performance capability in marine knowledge creation;

(d) An enabling environment for marine innovation.

3. The Selection Principles of Indicators

(a) An evaluation approach reflects a sustainable marine development strategy.

Consideration should be given to the holistic development environment of marine innovation as well as the indicators such as sustainability of economic development and knowledge achievement. At the same time attention should be given to time trend of indexes.

(b) The data used in the report all come from authoritative sources.

The basic data must come from national official statistics and surveys which are universally recognized. The statistics are collected through formal channels, on a timely and regular basis, to ensure accuracy, authoritativeness, and continuity.

(c) Indexes should be scientific, realistic and expandable.

The marine innovation index has a close logical relation with every sub-index. Every indicator of the sub-index should be scientific, objective and realistic to reduce the possibility of any artificial synthesis. Each indicator has a unique symbolic meaning at the macro level, defined in relatively broad terms rather than corresponding to the narrow scope of data, in order to facilitate the expansion and adjustment of the indicator system.

(d) An evaluation system should consider China's marine regional characteristics.

Index selection is mainly based on relative index, with consideration given to the various characteristics of output efficiency of marine innovation input, the scale of innovation activities and the width of innovation fields in different regions.

(e) Historical analysis and country-by-country comparison are both employed.

There are both review analyses of historical trends and comparisons between coastal areas, economic zones, economic circles and country-by-country analysis.

4. The Establishment of Index System

Innovation is an entire process from the introducing of innovation concepts to R&D and from knowledge output to commercial application, thus transforming knowledge into economic profits. Marine innovation capability is manifested in the whole process of production, circulation and transformation into economic profits of marine S&T knowledge. We should construct an indicator system for the evaluation of national marine innovation capability starting with the main links of the whole innovation chain, including marine innovation environment, innovation resources input, knowledge creation and application, and performance influence.

The report has adopted the evaluation method of composite index. From the perspective of innovation process, we have selected five sub-indexes, namely marine innovation resources, marine knowledge creation, marine enterprise innovation, marine innovation performance and marine innovation environment. Following the principles of indicator selection, we have developed an index system for evaluating national marine innovation index with 25 positive indicators (See Attached Table 1-1). Then, we have conducted comprehensive analysis, comparison and assessment of the marine innovation capability of our country based on national marine comprehensive innovation index and its supporting indicator system.

Marine innovation resources reflect a country's input in national marine innovation activities, the supply of innovation talents and investment in innovation infrastructure that innovation relies on. Innovation input is a prerequisite for marine innovation activities of a country including the funds investment in science and technology and resources of talents.

Marine knowledge creation reflects a country's capability in terms of marine scientific research output and knowledge dissemination. Marine knowledge creation is of various forms with multifaceted benefits. The report touches upon knowledge accumulation benefits of marine innovation from the perspectives of marine invention patents and scientific papers.

Marine enterprise innovation reflects the innovation capability of a country's marine enterprises. The indicator selection of marine enterprise innovation sub-index is based on the efficiency and effectiveness of marine enterprise innovation.

Marine innovation performance shows the effect and influence of a country's marine innovation activities. The indicator selection of marine innovation performance sub-index is based on marine innovation efficiency and effectiveness.

Marine innovation environment is the external environment underpinning a country's marine innovation activities, including related marine system innovation and environment innovation. Of those, the main body of marine innovation system is government and related departments, which is manifested in government's support to innovation policy, and capital support to innovation as well as the management of intellectual property rights. Environment innovation mainly refers to innovation in aspects of allocation capability, infrastructure, fundamental economic level, finance and cultural environment, etc.

Attached Table 1-1 Indicator System of National Marine Innovation Index

Comprehensive index	Sub-index	Indicator
National marine innovation index (A)	Marine innovation resources (B_1)	1. Input degree of R&D funds (C_1)
		2. Input degree of R&D personnel (C_2)
		3. Proportion of staff with intermediate and senior professional titles in total S&T personnel (C_3)
		4. Proportion of S&T staff in total personnel of marine scientific research institutions (C_4)
		5. Number of projects undertaken per 10 thousand scientific research personnel (C_5)
	Marine knowledge creation (B_2)	6. Number of invention patent applications per 100 million US dollars of economic output (C_6)
		7. Number of invention patent grants per 10 thousand R&D personnel (C_7)
		8. S&T works published in the current year (C_8)
		9. Number of S&T papers published per 10 thousand scientific research personnel (C_9)
		10. Proportion of papers published abroad in total articles (C_{10})
	Marine enterprise innovation (B_3)	11. Proportion of enterprise invention patent grants in total invention patent grants (C_{11})
		12. Ratio of enterprise R&D funds to the added value of major marine industries (C_{12})
		13. Number of invention patent grants per 10 thousand R&D personnel (C_{13})
		14. Independent rate of marine comprehensive technology (C_{14})
		15. Proportion of enterprise R&D personnel in total R&D personnel (C_{15})
	Marine innovation performance (B_4)	16. Transformation rate of marine S&T achievements (C_{16})
		17. Contribution rate of marine S&T progress (C_{17})
		18. Marine labor productivity (C_{18})
		19. Proportion of management service of science and research education in total gross ocean production (C_{19})
		20. Marine economic output per unit of energy consumption (C_{20})
		21. Proportion of gross ocean product in gross domestic product (C_{21})
	Marine innovation environment (B_5)	22. Per capita gross ocean product in coastal areas (C_{22})
		23. Proportion of equipment procurement cost in R&D funds (C_{23})
		24. Proportion of government funds in the total S&T funds for marine scientific research institutes (C_{24})
		25. Number of fresh graduates marine study with associate degree or above (C_{25})

Appendix 2: Definition of Indicators for National Marine Innovation Index

C1. Input Strength of R&D Funds

The indicator refers to the proportion of R&D funds of marine scientific research institutes in national gross ocean product and the input strength indicator of national marine R&D funds reflects the extent of a country's investment in marine innovation.

C2. Input Strength of R&D Personnel

The indicator, which refers to the R&D personnel per 10 thousand marine employees, reflects the strength of a country's investment in human resources for innovation.

C3. Proportion of Staff with Intermediate and Senior Professional Titles in Total S&T Personnel

The indicator, which refers to the proportion of researchers with intermediate and senior professional titles in marine scientific research institutes, reflects the power of highly skilled talents in a country's marine S&T activities.

C4. Proportion of S&T Staff in Total Personnel of Marine Scientific Research Institutes

The indicator, which refers to the proportion of S&T researchers in marine scientific research institutes, reflects a country's degree of commitment to scientific research manpower in marine innovation activities.

C5. Number of Projects Undertaken per 10 Thousand Scientific Research Personnel

The indicator, which refers to the number of domestic projects per 10 thousand scientific researchers, reflects the extent of innovation activities in which scientific researchers in marine field are involved.

C6. Number of Invention Patent Applications per 100 Million US Dollars of Economic Output

The indicator, which refers to the quotient of invention patent applications divided by gross ocean product (measured in 100 million dollars, calculated according to exchange rates), reflects the technical outputs relative to economic outputs and the level of support for a country's marine innovation activities.

Invention patent has the highest technological content and value among the three types of patent (invention patent, utility model patent and design patent). The number of invention patent applications can reflect a country's level of marine innovation activities and independent innovation capability.

C7. Number of Invention Patent Grants per 10 Thousand R&D Personnel

The indicator, which refers to the number of domestic invention patent grants per 10 thousand R&D personnel, reflects a country's independent innovation capability and technological innovation capability.

C8. S&T Works Published in the Current Year

The indicator refers to scientific works, college textbooks and popular science books published by authorized publishing houses. Only books with the institution's scientific researchers as the first author are included. Books of the same title are counted as one work regardless of circulation. This indicator reflects a country's output capability in marine science and research.

C9. Number of S&T Papers Published per 10 Thousand Scientific Research Personnel

The indicator reflects the efficiency of scientific research output.

C10. Proportion of Papers Published Abroad in Total Articles

The indicator, which refers to the proportion of papers published abroad in total S&T articles of a country, reflects the internationalization level of related research of scientific papers.

C11.Proportion of Enterprise Invention Patent Grants in Total Invention Patent Grants

The indicator reflects the contribution degree of marine enterprise patent invention to patent invention in marine field.

C12. Ratio of Enterprise R&D Funds to the Added Value of Major Marine Industries

The indicator reflects the input strength of marine enterprise innovation funds.

C13.The Number of Invention Patent Grants per 10 Thousand R&D Personnel

The indicator, which refers to the invention patent grants per 10 thousands enterprise R&D personnel, reflects independent innovation and technological innovation capabilities of a country's marine enterprises.

C14. Independent Rate of Marine Comprehensive Technology

The indicator refers to the average value between the proportion of technical revenues from enterprises in total technical income and the proportion of domestic patent grants in patent grants of the current years with regard to the revenues from regular funds of a country's marine scientific research institute, which reflects technical self-sufficiency of a country's marine industries.

C15. Proportion of Enterprise R&D Personnel in Total R&D Personnel

The indicator, which refers to the proportion of enterprise R&D personnel in total R&D personnel of a country's marine scientific research institutes, reflects the capability and standard of a country's marine enterprise R&D personnel input.

C16. Transformation Rate of Marine S&T Achievements

The indicator, which measures the transformation rate from marine S&T innovation achievements into commercial products, refers to the proportion of a series of activities which transform marine S&T achievements with practical value as a result of scientific research and technological development into new products, new techniques, new materials and new industry in the total S&T achievements in order to increase level of productivity. These activities include follow-up test, development, application and promotion that the transformation needs.

C17. Contribution Rate of Marine S&T Developments

The definition of this indicator is based on the definition of marine S&T growth rate. It refers to the proportion of marine S&T growth rate in total marine economic growth rate in various industries of marine economy. Marine S&T growth rate also refers to the growth of other factors, when humans use marine resources and marine space for social production, exchange, allocation and consumption, but exclusive of production factors such as capital and labor. It specifically refers to the increase of equipment technology level as a result of technological innovation, diffusion, transfer and import and the improvement of techniques, enhancement of employee quality and strengthening of management and decision-making capability.

C18. Marine Labor Productivity

The indicator, which refers to per capita gross ocean product produced by employees engaged in marine-related jobs, reflects the role of marine innovation activities in marine economic output.

C19. Proportion of Management Service of Science and Research Education in Gross Ocean Product

The indicator reflects the contribution of marine scientific research, education, management, service and other activities to the marine economy.

C20. Marine Economic Output per Unit Energy Consumption

The indicator, which refers to GOP per 10,000 standard-coal consumption, measures the reduction of energy consumption due to marine innovation and reflects a country's intensification level of marine economic growth.

C21. Proportion of Gross Ocean Product in Gross Domestic Product

The indicator reflects the contribution of marine economy to national economy and measures the supportive impact of marine innovation on marine economy.

C22. Per Capita Gross Ocean Product in Coastal Areas

The indicator, to some degree, reflects the living standards in coastal areas, and measures the growth status of marine productivity and the external environment for marine innovation activities.

C23. Proportion of Equipment Procurement Cost in R&D Funds

The indicator, which refers to the proportion of equipment procurement cost in R&D funds in marine scientific research institutes, shows the hardware requirements that marine innovation needs and reflects the physical environment of marine innovation to some degree.

C24. Proportion of Government Funds in the Total S&T Funds for Marine Scientific Research Institutes

The indicator reflects the supportive effect that the government's investment has on marine innovation and the system environment for marine innovation.

C25. Number of Fresh Graduates of marine study with Associate Degree or Above

The indicator reflects a country's capability in the training and supply of marine S&T human resources.

Appendix 3: Evaluation Method of National Marine Innovation Index

National marine innovation index is calculated through the use of benchmarking, a popular calculation method adopted by the *IWD World Competitiveness Yearbook*. Benchmarking is a methodology widely used in the world today. It works in the following way: first, a benchmark value is set, then the subjects of evaluation will be measured against the benchmark value to identify the gap among each other and show the results of ranking.

We adopt the indicators of the evaluation indicator system of marine innovation and use the indicator data between 2001 and 2015 to calculate the scores of China's marine innovation index and sub-index respectively in years afterwards. A comparison of the scores obtained with the score of the base year in this way can capture the growth trend of national marine innovation index.

1. Standardized Processing of the Original Data

With 2001 as the base year and 100 as the base value, the standardized processing of original data of the 25 indicators in the indicator system of marine innovation index is as follows:

$$C_j^t = \frac{100x_j^t}{x_j^1}$$

In the equation, j =1~25 is the sequence number of the indicator; t =1~15 is the sequence number of the reference year (2001-2015); x_j^t is the original data for the reference year's indicators (x_j^1 is the original data of 2001 indicators); C_j^t is the value after standardized processing of the indicators.

2. Calculation of National Marine Innovation Sub-index

The scores for the sub-index are calculated based on equal weight[1] (the same as below) :

When $i=1$, $B_1^t = \sum_{j=1}^{5} \beta_1 C_j^t$, $\beta_1 = \frac{1}{5}$;

When $i=2$, $B_2^t = \sum_{j=6}^{10} \beta_2 C_j^t$, $\beta_2 = \frac{1}{5}$;

When $i=3$, $B_3^t = \sum_{j=11}^{15} \beta_3 C_j^t$, $\beta_3 = \frac{1}{5}$;

When $i=4$, $B_4^t = \sum_{j=16}^{21} \beta_4 C_j^t$, $\beta_4 = \frac{1}{6}$;

1 Equal weight is obtained based on weight selection method of National Innovation Index Report 2014

$$\text{When } i=5,\ B_5^t=\sum_{j=22}^{25}\beta_5 C_j^t,\ \beta_5=\frac{1}{4}.$$

In the equations, i = 1~5; t = 1~14, B_1^t, B_2^t, B_3^t, B_4^t, B_5^t respectively represents the score of marine innovation resources sub-index, marine knowledge creation sub-index, marine enterprise innovation sub-index, marine innovation performance sub-index and marine innovation environment sub-index.

3. Calculation of National Marine Innovation Index

The scores of national marine innovation index are calculated based on equal weight (ibid):

$$A^t=\sum_{i=1}^{5}\varpi B_i^t$$

In the equation, i = 1~5; t = 1~14; ϖ is the equal weight (equal weight = $\frac{1}{5}$) ; A^t is the score of national marine innovation index in different years.

Appendix 4: Evaluation Method of Regional Marine Innovation Index

1. Explanation on Indicator System of Regional Marine Innovation Index

The indicator system of regional marine innovation index is composed of sub-indexes of marine innovation resources, marine knowledge creation, marine innovation performance and marine innovation environment. Compared with indicator system of national innovation index, it lacks marine enterprise innovation sub-index. Meanwhile, compared with the sub-index of national marine innovation performance, the sub-index of regional marine innovation performance is short of two indicators of "contribution rate of marine S&T developments" and "transformation rate of marine S&T achievement".

2. Normalized Processing of Original Data

We have conducted normalized processing of the original value of the 18 indicators of 2015. Normalized processing is used for the purpose of removing the discrepancies in measurement unit, magnitude order of indicator value and form of relative number, thereby solving the comparability issue of data indicator and putting indicators in the same magnitude order for the convenience of comprehensive analysis and contrast in multi-indicator comprehensive evaluation.

The indicator data is processed based on linear normalization:

$$c_j = \frac{y_j - \min y_j}{\max y_j - \min y_j}$$

In the equation, i = 1~11 is the sequence numbers of eleven coastal provinces (cities) in Chinese mainland; j = 1~18 is the sequence numbers of the indicator; y_j is the original data value of the indicator; c_j is the normalized data value.

3. Calculation of the Sub-index of Regional Marine Innovation

The score of the sub-indicators of regional marine innovation environment:

$$b_1 = 100 \times \sum_{j=1}^{5} \varphi_1 c_j,\ \varphi_1 = \frac{1}{5}$$

The score of sub-indicators of regional marine knowledge creation:

$$b_2 = 100 \times \sum_{j=6}^{10} \varphi_2 c_j,\ \varphi_2 = \frac{1}{5}$$

The score of sub-indicators of regional marine innovation performance:

$$b_3 = 100 \times \sum_{j=11}^{14} \varphi_3 c_j,\ \varphi_3 = \frac{1}{4}$$

The score of sub-indicators of regional marine innovation environment:

$$b_4 = 100 \times \sum_{j=15}^{18} \varphi_4 c_j,\ \varphi_4 = \frac{1}{4}$$

In the equation, j =1~18 ; b_1, b_2, b_3, b_4 respectively represents the sub-indicator scores of regional marine innovation resources, regional marine knowledge creation, regional marine innovation performance and regional marine innovation environment.

4. Calculation of Regional Marine Innovation Index

The scores of regional marine innovation index are calculated based on equal weight (the same as the national marine innovation index):

$$a = \frac{1}{4}(b_1 + b_2 + b_3 + b_4)$$

In the equation, a is the score of the regional marine innovation index.

Appendix 5: Analysis on the Contribution Rate of Marine S&T Progress

As an important indicator to measure the contribution of marine S&T progress to the growth of marine economy, the contribution rate of marine S&T progress has attracted more and more attention, and the research done by domestic scholars on this indicator has been increasing day by day. However, due to different and inadequate understanding, scholars tend to be biased when applying this indicator, thus affecting public's objective knowledge and comprehensive understanding of marine S&T innovation capability. In view of this problem, this study elaborates and differentiates the origin, connotation and other aspects as follows.

1. The Origin and Significance of Contribution Rate of Marine S&T Progress

The contribution rate of marine S&T progress is derived from the contribution rate of S&T progress and the principle and measurement methods of the contribution rate of S&T progress come from production function and the *Solow Growth Equation* (also known as *Solow Residual Method*) which was improved on the basis of production function. A large-scale measurement of the contribution rate of S&T progress began in 1992, and State Development Planning Commission (SDPC) issued the *"Circular on Measuring S&T Progress in Economic Growth"* (Science and Technology [1992] No. 2525), creating a new upsurge in studying the contribution rate of S&T progress in various fields among experts and scholars. Since then, the contribution rate of S&T progress which acts as an important indicator of innovation fields has increasingly appeared in national plans including the "*National Medium and Long-term Program for Science and Technology Development*", "*National '11th Five-Year Plan' for Science and Technology Development*", "*National'12th Five-Year Plan' for Science and Technology Development*". The indicator "contribution rate of S&T progress" is also newly added to the recently released "*The '13th Five-Year Plan' for Economic and Social Development*".

The contribution rate of S&T progress is also of great significance to the ocean. After entering the 21st century, China's marine strategic position has been gradually improved. At the same time, marine science and technology have become the core element and an important supporting force to promote marine economic development. To make a quantitative evaluation of the function of marine S&T progress in marine economic growth will play supporting and guiding roles in the formulation of marine S&T development strategy and related marine S&T policy. At present, the contribution rate of marine S&T progress has been applied in the "*Outline of Marine development by means of Science and Technology (2008-2015)*", "*National '12th Five-Year Plan' for Marine Economy Development*", "*National'12th Five-Year Plan' for Marine Industry Development*" and "*National Marine Development Plan*". *"The Overall Plan for Marine Power Strategy and S&T Innovation"* which is being compiled

also uses the indicator.

2. Analysis on the Connotation of Contribution Rate of Marine S&T Progress

What is the contribution rate of marine S&T progress? What's the relationship between contribution rate of marine S&T progress and total factor productivity? Can contribution rate of marine S&T progress maintain the growth trend as always? Is contribution rate of marine S&T progress appropriate for regional evaluation? Here are the elaborations one by one to these frequently asked questions.

(a) What is the contribution rate of marine S&T progress?

What's the connotation of contribution rate of marine S&T progress? Firstly, we must clarify the contribution of S&T progress to economic growth. Theoretically, the contribution made by S&T progress to economic growth is revealed as an expansion of reproduction, the principle of which can be understood as the process of all factors working together when a certain amount of productive factors are combined to furbish more products (use value), which can be specifically summarized in the following several aspects: improving the technical level of equipment, modifying technical processes, enhancing the quality of the workforce and increasing management decision-making power i.e. those factors affecting economic growth excluding the contribution of capital and the labor force to that economic growth. According to macroeconomics, apart from the input of labor and capital, only the improvement of technological levels can contribute to medium- and long-term economic growth. Therefore, the contribution of medium- and long-term comprehensive factors can be called the contribution of S&T progress.

In the context of the ocean, the contribution rate of marine S&T progress should be defined on the basis of the definition of the growth rate of marine S&T progress. The growth rate of marine science and technology refers to the contribution to marine economic growth made by human beings when people are making use of marine resources and marine space to undertake different kinds of production and service activities and are using marine resources in activities such as social production, exchange, allocation and consumption (excluding the contribution to marine economic growth made by the increase of production factors such as capital and labor force). The share of the growth rate that marine S&T progress accounts for in the growth rate of the marine economy is the contribution rate of marine S&T progress (percentage). Its economic implication refers to the combination of a number of production factors to gain more production (use value). It can also be understood as the share that the increase of other factors accounts for, excluding the fixed factors such as capital and labor force, in the growth of marine economy.

(b) What's the relationship between the contribution rate of marine S&T progress and total factor productivity?

Total Factor Productivity (TFP) is an international general indicator, which comes from technological progress, organizational innovation, specialization and production innovation. The rate at which the output growth rate exceeds the factor input growth rate is usually called the TFP growth rate. The connotation of S&T progress in the contribution rate of marine S&T progress is the same as all the elements in the TFP after deducting capital input and labor input. The contribution rate of marine S&T progress can also be expressed as the ratio of marine TFP to marine economic growth rate.

Zhu Guangyao, CPPCC National Committee member and vice minister of the Ministry of Finance, proposed on March 7, 2015 during the CPPCC discussions that the newly added contribution rate of S&T progress in the "*National '13th Five-Year Plan' for Economic and Social Development (Draft)*" is crucially important. He believes that this indicator is conducive to improving China's TFP, which indirectly reflects the correlation between the contribution rate of S&T progress and TFP. In fact, the economic connotation of the contribution rate of S&T progress is the contribution rate of TFP, but from a long time scale, economists believe that only S&T progress can continuously promote economic growth. In other words, the measurement of the contribution rate of S&T progress is based on TFP at long-time scale. In practical application, in order to minimize the measurement error, the author suggests that the measurement time length used to calculate the contribution rate of national marine S&T progress should be more than 10 years, or at least 5 years.

(c) Can contribution rate of marine S&T progress maintain the growth trend continuously?

The contribution rate of marine S&T progress is a relative indicator and its numerical value will not continue to increase due to the following two reasons.

On the one hand, the contribution of marine sci-tech to economic growth has the characteristics of hysteresis and periodicity, so the indicator value does not always increase, but tends to fluctuate. Economic cycle, natural disasters, policy changes, and so forth can affect the indicator value. TFPs of the United States, Europe and other developed countries are also constantly fluctuating.

On the other hand, the indicator value of the contribution rate of marine S&T progress is difficult to continue to increase significantly when the marine S&T development reaches certain level. From the historical point of view, the impact of marine S&T progress on marine economic growth should be continuous and cumulative. In other words, the marine S&T progress at a certain point will continue to affect all activities afterwards. Marine economic growth during any given period of time is attributed to the contribution of marine S&T progress within this period and all marine S&T progress before that period.

(d) Is the contribution rate of marine S&T progress appropriate for regional evaluation?

It is of little significance to use the contribution rate of marine S&T progress to make interregional horizontal comparison between cities and counties.

By definition, the contribution rate of S&T progress is the share of economic growth brought about by S&T progress and other factors with the exclusion of capital and labor factors, which is used to measure the role that technological progress plays in increments rather than aggregates. Therefore, this indicator is of little significance for horizontal comparison but more suitable for vertical comparison of a country or region. For example, social economy and S&T development of the eastern coastal areas of China are relatively advanced, if we make measurement with present stage as base period, the numerical value of the contribution rate of S&T progress of the region will be smaller than that of backward inland regions.

From a theoretical point of view, the general level of science and technology within a country is relatively close, because there is no technical barrier between countries. Therefore, it is not recommended to compare the contribution rate of S&T progress between cities and countries horizontally.

At different stages of economic development, the contribution of S&T progress to economic growth is different. It is not recommended to carry out horizontal comparisons of different regions at different stages. If necessary, it is better to make the comparison along with other indicators such as economic development level and developing degree of society.

3. Conclusion

With improved demands for quantitative indicators of various marine plans, the contribution rate of marine S&T progress will gain more attention. We should be down to earth and strengthen research on data acquisition, method selection and measurement process so as to minimize errors and give full play to the practical application and guiding value of the contribution rate of marine S&T progress.

Appendix 6: Calculation Methods of Contribution Rate of Marine S&T Progress

Currently, the method of calculating the contribution rate of S&T progress that is widely and commonly used is *Solow Residual Method*. It is also a method commonly used by National Development and Reform Commission (NDRC, formerly the National Planning Commission), the National Bureau of Statistics and the Ministry of Science and Technology, etc.

With Cobb-Douglas Production Function as a base model, *Solow Residual Method* indicates that in addition to capital growth rate, labor growth rate and the weight of the relative effect that capital and labor have on income growth, economic growth also depends on technological progress. Here, it distinguishes "growth effect" caused by the increase of quantities of different factors from the "level effect" of economic growth as a result of the improvement of technological levels, systematically explaining the reasons for economic growth.

Marine economy involves many industries and sectors. In order to comprehensively reflect the contribution of S&T progress in various marine industries to the overall increase of marine economy, calculation and measurement across the board need to be conducted to all types of marine industries. Then according to the proportion of total economic output of various industries in marine economy, we calculate the growth rate of marine science and technology by weighting the S&T progress of various industries during the measuring phase of growth rate. Based on that, the contribution rate of marine science and technology will be calculated.

The industries and sectors that the contribution rate of marine S&T progress involves based on its theoretical connotation and characteristics are as follows: production and service about products obtained directly from the ocean; one-time processing production and service about products obtained directly from the ocean; production and service about products which are directly applied to the ocean; production and service using seawater and ocean space as the basic elements of production process. Among them, other services and management scopes such as marine scientific research, education and technology are not appropriate to be calculated by the contribution rate of marine S&T progress.

In combination with our country's marine S&T characteristics and through industry weighting to the output growth rate, capital growth rate and labor growth rate of eight marine industries, the basic formula to calculate the contribution rate of marine S&T progress is constructed as follows:

With "i" industry (i = 1, 2, 3, ..., 8) representing marine culture, marine fishing, marine salt industry, marine shipping, marine oil, marine natural gas, marine transportation, and coastal tourism respectively:

$y_i(t)$ indicates the output growth rate of i industry during t period, among which $t \in [t_1, t_2]$;

$k_i(t)$ and $l_i(t)$ indicate the growth rate of capital and labor input within t period respectively, among which $t \in [t_1, t_2]$;

γ_i indicates the weight of i industry in overall marine industries.

k_i, l_i, y_i indicate the average value between t_1 and t_2 at the research interval of $k_i(t)$, $l_i(t)$, $y_i(t)$ respectively, namely:

$$k_i = \frac{\sum_{t=t_1}^{t_2} k_i(t)}{n},\ l_i = \frac{\sum_{t=t_1}^{t_2} l_i(t)}{n},\ y_i = \frac{\sum_{t=t_1}^{t_2} y_i(t)}{n}$$, among which $n = t_2 - t_1$

k, l, y indicate weighted average of k_i, l_i, y_i, namely:

$$k = \sum_{i=1}^{8} k_i\gamma_i, l = \sum_{i=1}^{8} l_i\gamma_i, y = \sum_{i=1}^{8} y_i\gamma_i .$$

The formula can be obtained:

$$A = 1 - \frac{\alpha k}{y} - \frac{\beta l}{y} = 1 - \frac{\alpha\sum_{i=1}^{8} k_i\gamma_i}{\sum_{i=1}^{8} y_i\gamma_i} - \frac{\beta\sum_{i=1}^{8} l_i\gamma_i}{\sum_{i=1}^{8} y_i\gamma_i}$$

$$= 1 - \frac{\alpha\sum_{i=1}^{8}\frac{\sum_{i=t1}^{t_2} k_i(t)}{n}}{\sum_{i=1}^{8}\frac{\sum_{i=t_1}^{t_2} y_i(t)}{n}\gamma_i} - \frac{\beta\sum_{i=1}^{8}\frac{\sum_{i=t1}^{t_2} l_i(t)}{n}}{\sum_{i=1}^{8}\frac{\sum_{i=t_1}^{t_2} y_i(t)}{n}\gamma_i}$$

Among which, A is the contribution rate of marine S&T progress within the period of research; α and β indicate marine industry capital and elasticity coefficient of labor respectively.

With regard to the duration of indicator selection, measurement time length used to calculate the contribution rate of national marine S&T progress should be more than ten years, or at least five years because the influence of marine science and technology on marine economy is long term. Considering the actual need of marine management and time limit of marine data, the study uses the average value of five years data when calculating the indicators of "*the 11th five-year plan*" period and doing short-term forecast to the indicators of "*the 12th five-year plan*". The average value of 10 years is used when doing other calculations and long-term forecast (depending on the time length from 2006 to 2015).

With regard to marine industry selection, according to "*China Marine Statistical Yearbook 2016*" (the latest data), the twelve major marine industries in 2015 are as follows: marine fishery (16.25%), marine oil and gas industry (3.51%), marine mining (0.25%), marine salt industry (0.26%), marine shipping industry (5.38%), marine chemistry industry (3.68%), marine bio pharmaceutics industry (1.13%), marine engineering construction industry (7.81%), marine electric power industry (0.43%), seawater utilization industry (0.05%), marine transportation industry (20.68%) and coastal tourism industry (40.59%) (See Attached Table 6-1). After preliminary screening and feasibility analysis, eight industries are calculated and measured with supporting data. They are marine culture, marine fishing, marine salt industry, marine shipping, marine oil, marine natural gas, marine transportation, and coastal tourism. The total production value of the above-mentioned eight industries accounts for 86.65% of the total production value of the major marine industries, which can effectively reflect the development of China's marine economy.

Attached Table 6-1 Added Value of the Major Marine Industries of China in 2015

Major marine industries	Added value(billion Yuan)	Proportion (%)
Total	**26 791.3**	—
marine fishery	4 352.4	16.25%
marine oil and gas industry	939.3	3.51%
marine mining	67.1	0.25%
marine salt industry	68.6	0.26%
marine shipping industry	1 440.5	5.38%
marine chemistry industry	985.0	3.68%
marine bio pharmaceutics industry	301.6	1.13%
marine engineering construction industry	2 092.1	7.81%
marine electric power industry	116.0	0.43%
seawater utilization industry	14.0	0.05%
marine transportation industry	5 540.8	20.68%
coastal tourism industry	10 874.1	40.59%

With regard to the elasticity coefficient, the elasticity coefficient of capital and labor output can be determined by adopting experience estimation method, ratio method and regression method when calculating the contribution rate of marine S&T progress. Experience estimation method means using the coefficient calculated by other authoritative experts as reference; the principle of ratio method is to make use of the data related to the amount of capital input and labor input to calculate their ratio; regression method refers to the use of constrained production function model (namely $\alpha+\beta=1$) or unconstrained production function model to estimate two coefficients after locating relevant numerical values according to measurement method (namely using the least square method to regress). The method adopted by the study is: $\alpha=0.3$, $\beta=0.7$.

With regard to the weightings, the weight values (See Attached Table 6-2) of the eight marine industries are determined according to their output values during "*the 12th five-year plan*" period mentioned in "*China Marine Statistical Yearbook*".

Attached Table 6-2 Weight Values of Each Industry

Industry	Weight value	Industry	Weight value
marine culture industry	0.1096	marine oil industry	0.0709
marine fishing industry	0.0810	marine natural gas industry	0.0045
marine salt industry	0.0033	marine transportation industry	0.2489
marine shipping industry	0.0664	coastal tourism	0.4154

With regard to data sources, all the indicator data representing marine industry value, capital and labor adopted by this study are from "*China Marine Statistical Yearbook*" of corresponding year (See Attached Table 6-3). In terms of the basic data, the continuous data that can be calculated are the output value of marine industries, capital and labor from 1996 to 2015 (trend fitting interpolation are used to deal with individual missing data).

Attached Table 6-3 Indicators of the Output, Capital and labor of the Eight Industries

Eight industries	Output indicator	Capital indicator	Labor indicator
marine culture industry	output of marine culture industry	area of marine culture	number of the employed of marine culture and relevant industries
marine fishing	output of marine fishing	total tons of ships working on the sea	number of the employed of marine fishing and relevant industries
marine salt industry	sea salt production of coastal areas	production area of marine salt industry	number of the employed of marine salt industry
marine shipping	added value of marine shipping	completion amounts of ship building of coastal areas	number of the employed of marine shipping
marine oil industry	marine crude oil production of coastal areas	offshore oil wells	number of the employed of marine oil and natural gas industry
marine natural gas industry	marine natural gas production of coastal areas	offshore gas wells	number of the employed of marine oil and natural gas industry
marine transportation industry	added value of marine transportation industry	number of berths used for production in ports above designated size of coastal areas	number of the employed of marine transportation industry
coastal tourism	added value of coastal tourism	total number of travel agencies of coastal areas	number of the employed of coastal tourism

To locate the benchmark data of each industry within the formula measuring the contribution rate of marine S&T progress, we can obtain the average value of the contribution rate of marine S&T progress. After calibration and verification, the average values of the contribution rate of marine S&T progress in China during "*the 11th Five-Year Plan*" period and the period from 2006 and 2015 were 54.40% and 64.2% respectively (See Attached Table 6-4).

Attached Table 6-4 Measurement Value of Contribution Rate of Marine S&T Progress

Year	Output growth rate (%)	Capital growth rate (%)	Labor growth rate (%)	Contribution rate of marine S&T progress (E) (%)
2006—2010	12.86	10.10	4.05	54.4
2006—2015	10.97	6.74	2.72	64.2

As can be seen from the Attached Table 6-4, the contribution rates of marine S&T progress of China during "*the 11th Five-Year Pla*n" period and the period from 2006 to 2015 were 54.4% and 64.2% respectively. The significant increase is attributed to the great importance attached to marine science and technology in recent years. The "*Outline of the National 12th 'Five-Year' Plan for Marine Scientific and Technological Development*" points out that the marine S&T innovation capability and S&T supporting capability must be enhanced to promote rapid, sustainable and healthy development of our marine economy, effectively transform the growth mode of marine economy and make science and technology an important force to underpin and lead marine innovation development.

Appendix 7: Calculation Methods of Transformation Rate of Marine S&T Achievements

The definition of marine S&T achievements transformation rate is derived from S&T achievements transformation rate. With regard to the studies on the transformation rate of S&T achievements, foreign scholars rarely use the term "S&T achievements transformation" directly, but replace it with "integration of technology and economy", "technology innovation", "technology transformation", "technology promotion", "technology diffusion" or "technology transfer". In addition, there is no statistical analysis or evaluation done to transformation situations of S&T achievement of the whole society in foreign countries.

Domestically, scholars in various fields are divided in the definitions of S&T achievements transformation, which can be reduced to the following three points:

According to the first viewpoint, the transformation rate of S&T achievements refers to the ratio of transformed S&T achievements in all applied technological achievements. Scholars believe that "the transformed S&T achievements" do not refer to all S&T achievements that have been "transformed". We should investigate the acceptance of the market to the technological achievements or their direct or indirect benefits when technology achievements are applied to production. If the applied technology achievements can be successfully transformed into commodities and achieve economies of scale, then the applied technology achievements have realized transformation.

According to the second viewpoint, the transformation rate of S&T achievements refers to the ratio of transformed S&T achievements in all S&T achievements. Scholars believe that although most of the fundamental theoretical achievements and some soft science achievements cannot be directly applied to the actual production and the achievement transformation is of low-level quantification, they are still able to promote the progress of science and technology and the adjustment and optimization of industrial structure to a certain extent. Therefore, it is recommended that the transformation of basic theoretical achievements and soft science achievements should be incorporated into S&T achievement transformation.

According to the third viewpoint, from the perspective of management, the S&T achievement transformation rate should show the ratio of S&T achievement in all research subjects.

The second viewpoint should not be adopted because the fundamental research achievements and soft science research achievements in marine field can hardly be directly applied to the actual production, which makes it difficult to realize the transformation of S&T achievements. As to the third viewpoint, "S&T achievements" and "research subjects" in the definition come from two sets of different marine statistical data with the former one deriving from the marine S&T statistical data while the latter one from the marine S&T achievements statistical data, so this viewpoint cannot accurately reflect the actual transformation of marine S&T achievements.

Therefore, this report adopts the first viewpoint with the definition of transformation rate of marine S&T achievements as follows:

The transformation rate of marine S&T achievements refers to the percentage of marine S&T achievements of marine-related enterprises within a certain period in the total application of marine S&T achievements. These marine S&T achievements can realize self-transformation or be transformed into production and they must be at the application or production stage and have reached maturity in application. According to the definition of marine S&T achievement transformation rate, the formula for marine S&T achievement transformation rate can be determined as follows:

Transformation rate of marine S&T achievements = maturely applied marine S&T achievements / all achievements of marine S&T applied technology × 100%

As the transformation of marine S&T achievements is a long-term process, the longer the coverage period is, the more realistic the indicators are when measuring and calculating the transformation rate of marine S&T achievements.

It should be noted that the transformation rate of marine S&T achievements discussed in this report refers to indicators in a narrow term. The "maturely applied marine S&T achievement" and "all achievements of marine S&T applied technology" in the formula come from registered data of marine S&T achievements. In the broad sense, marine scientific research projects, patents, papers, awards, standards, software and copyright all belong to marine S&T achievements, so it is difficult to get the statistics and there is overlapping between each other. It is extremely hard to identify and measure the process because all steps from the formation of marine S&T achievements, to the initial application, to the formation of products, until the scalization and industrialization stage, can be counted into marine S&T achievements transformation process.

The transformation rate of marine S&T achievements in our country between 2000 and 2015 was approximately 50.1% according to the calculation using standard formula for the transformation rate of marine S&T achievements, based on the statistics of marine S&T achievements.

Note: According to the S&T achievements registration form, the applied technology achievements can be divided into three stages. The initial stage refers to the research achievements of laboratory, small test at the early stages. Mid-term stage refers to the intermediate test (mid-term test) which can further improve the product, technique or production process before the new products, new technique and new production process are directly used for production; the prototype and the sample made for product design and getting the technical parameters needed for production; demonstration for wide promotion; phasic research achievements for mature application and wide promotion. Mature application stage refers to the achievements that have been applied officially (or can be officially) in industrial production, including large-scale promotion of agricultural technology, clinical application of medical health care and other achievements such as the flight model and design finalization of public security and military industry.

Appendix 8: Calculation Methods and Indicator System of Marine S&T Input-output Efficiency

1. Calculation Methods

With city as basic research unit, marine science and technology input-output efficiency as research subject, data of the marine scientific research institutes during the 2001 to 2015 period as the research base, and the DEA as the model, this chapter has done calculation and measurement to the input-output efficiency of marine-related cities and then explored the rules of it. On this basis, the chapter will do retrospective analysis and trend prediction to the input-output efficiency of marine science and technology from the *"10th Five-Year"* period to the *"13th Five-Year"* period in China, in order to provide data support in promoting the development of marine science and technology.

The Data Development Analysis (DEA) is a nonparametric statistics analysis put forward by A.Charnes and W.W. Cooper, which is widely used in the evaluation of the input-output efficiency among countries or regions. The advantages of DEA are as follows: (a) The weight of the input and output needn't to be determined in advance because of the simple structure of DEA; (b) The feature of units invariance enables it to employ different measurement units; (c) Comparative analysis and efficiency analysis can be done between the actual values and target values. The research uses the DEA model with variable returns to scale (VRS) and take the national marine-related cities as the decision making unit (DMU) of evaluating marine S&T input-output efficiency to calculate the overall efficiency, the pure technical efficiency and the scale efficiency of marine-related cities.

Suppose there are N DMUs and each DMU owns K inputs and L outputs, then set $x_{nk}(x_{nk}>0)$ as the kth resource input of the nth marine-related city, and $y_{nl}(y_{nl}>0)$ as the lth achievement output of the nth marine-related city. As for the nth coastal city, $\theta(0<\theta\leqslant 1)$ refers to the overall index of the marine S&T input-output efficiency, ε is the non-Archimedean infinitesimal quantity and $\lambda_n(\lambda_n\geqslant 0)$ is the weight variable, which are used in estimating the returns to scale of the marine-related cities. In addition, $s^-(s^-\geqslant 0)$ is a slack variable which refers to the input needed to be reduced when marine-related cities achieve the DEA effectiveness and $s^+(s^+\geqslant 0)$ is a residual variable which refers to the output needed to be added when marine-related cities achieve the DEA effectiveness. The models are as follows:

$$
\begin{cases}
\min(\theta - \varepsilon(\sum_{k=1}^{K} s^- + \sum_{l=1}^{L} s^+)) \\
s.t. \sum_{n=1}^{N} x_{nk}\lambda_n + s^- = \theta x_k^n \quad k=1, 2, ..., K \\
\sum_{j=1}^{n} y_{nl}\lambda_n - s^+ = y_l^n \quad l=1, 2, ..., L \\
\lambda_n \geqslant 0, n=1, ..., N
\end{cases}
$$

The overall efficiency of marine S&T input-output, pure technical efficiency and scale efficiency of the DMU can be calculated based on the models mentioned above, among which the overall efficiency refers to the comprehensive level of the marine S&T resource allocation and the utilization of the marine S&T resources, reflecting the effectiveness of the input and output of the DMU. The pure technical efficiency refers to the degree of the effective utilization of the marine S&T input elements by DMU and the scale efficiency refers to the disparity between the input scale and the optimal scale of the marine S&T resources.

2. Establishment of the Indicator System

Since the establishment of evaluating indicator system is the key in DEA evaluation model, truthful and objective sample data will be selected and scientific and rational indication system will be established to guarantee the accuracy and reliability of the evaluation results. among the existing researches on the evaluation of the science and technology input-output efficiency, most adopt the labor force and the capital as the input variables and academic papers, publications and patents as the output variables. According to the principle of "the fewer, the better" in terms of the number of the input-output efficiency indicators, numbers of the input indicators + numbers of the output indicators ⩽ 1/3 DMUs, and taking the accessibility of the data and the features of large input scale and long payback period of marine S&T development into a whole consideration, the evaluation indicator system of the research is established as follows:

Marine S&T input can reflect the basis and conditions of marine science and technology R&D as well as incubation in marine-related areas, including marine S&T capital investment, talents injection and numbers of marine-related scientific research institutes, which could provide strong driving force for marine S&T innovations. Moreover, science and technological expenditure refers to the actual expenses that marine-related units incurred in carrying our internal marine scientific research activities, while the science and technological personnel refers to the science and technology management personnel, the research related staff and the science and technological service personnel among staff of marine-related units.

Attached Table 8-1 Indicator System of Marine S&T Input-output Evaluation

First class indicator	Second class indicator	Third class indicator
Marine science and technology input	Capital input	Expenditures of Science and technology activities
		Numbers of marine-related scientific research institutes
	Labor force input	Personnel involved in S&T activities
Marine science and technology output	Science and technology output	Numbers of published S&T papers
		Numbers of released S&T publications
		Numbers of authorized S&T patents

Marine science and technology output reflects the transformation and utilization ability of the marine S&T resources in marine-related regions, including numbers of published S&T papers, numbers of published works and numbers of authorized patents. In addition, the driving force of marine S&T development to the marine economy will also be regarded as the range of marine S&T output; however, it fails to be incorporated into the indicator system for the lack of relevant data.

Appendix 9: Regional Classification Basis and Definition of Related Concepts

1. Coastal Provinces (cities)

Eleven provinces (cities) have coastlines, including Tianjin, Hebei, Liaoning, Shanghai, Jiangsu, Zhejiang, Fujian, Shandong, Guangdong, Guangxi and Hainan.

2. Marine Economic Zones

China has five marine economic zones, namely, Bohai Rim Economic Zone, the Yangtze River Delta Economic Zone, the west coast of the Taiwan Straits Economic Zone, the Pearl River Delta Economic Zone and the Beibu Gulf Rim Economic Zone. Coastal provinces (cities and districts) which are incorporated in the evaluation of the Bohai Rim Economic Zone are Liaoning, Hebei, Shandong and Tianjin; coastal provinces (cities, districts) which are incorporated in the evaluation of the Yangtze River Delta Economic Zone are Jiangsu, Shanghai and Zhejiang; Fujian is incorporated in the evaluation of the west coast of the Taiwan Straits Economic Zone; Guangdong is incorporated in the evaluation of the Pearl River Delta Economic Zone; coastal provinces (cities, districts) which are incorporated in the evaluation of the Beibu Gulf Rim Economic Zone are Guangxi and Hainan.

3. Marine Economic Circles

China boasts three marine economic circles including the northern marine economic circle, eastern marine economic circle and southern marine economic circle based on the "*the '12th Five-Year' Plan for National Marine Economic Development*". The northern marine economic circle comprises Liaodong Peninsula, the Bohai Bay and the coasts and sea areas of Shandong Peninsula, and the coastal provinces (cities and districts) which are incorporated in this evaluation include Tianjin, Hebei, Liaoning and Shandong; the eastern marine economic circle is composed of Jiangsu, Shanghai and the coasts and sea areas of Zhejiang, and the coastal provinces (cities, districts) which are incorporated in this evaluation include Jiangsu, Zhejiang and Shanghai; the southern marine economic circle comprises Fujian, the Pearl River Estuary and its two wings, the Beibu Gulf and the coasts and sea areas of Hainan Island, and the coastal provinces (cities, districts) which are incorporated in this evaluation include Fujian, Guangdong, Guangxi and Hainan.

4. Marine-related Cities

The marine-related cities in this study refer to cities with marine scientific research institutes and this concept is only limited to this study. According to the S&T statistics from the Ministry of Science and Technology, there are totally 59 marine-related cities nationwide from 2001 to 2015 which are shown in the

Attached Table 9-1. According to geographical space, marine-related regions can be divided into northern, eastern, southern, southwestern and mid-western parts, of which the northern part includes Beijing, Tianjin, Shandong, Liaoning and Heilongjiang, the eastern part includes Shanghai, Jiangsu, Zhejiang and Fujian, the southern part includes Guangdong and Hainan, southwestern coastal part includes Guangxi Zhuang Autonomous Region. In spite of relatively strong marine scientific research strength, parts of landlocked areas are included into marine-related areas of mid-west due to obvious disparities from Guangxi Zhuang Autonomous Region in the aspects of region position and data. Based on the principles above, northern parts have 23 marine-related cities, followed by eastern part having17, southern part having 13 and coastal areas of southwest and marine-related areas of mid-west both having 3.

Attached Table 9-1　List of Marine-related Cities

Provinces (cities, districts)	Marine-related cities				
Beijing	Beijing				
Tianjin	Tianjin				
Hebei	Shijiazhuang	Qinhuangdao			
Liaoning	Shenyang	Dalian	Fushun	Dandong	Jinzhou
	Yingkou	Liaoyang	Panjin	Tieling	Huludao
Heilongjiang	Haerbin				
Shanghai	Shanghai				
Jiangsu	Nanjing	Nantong	Lianyungang	Yancheng	Zhenjiang
Zhejiang	Hangzhou	Ningbo	Wenzhou	Shaoxing	Zhoushan
	Taizhou				
Fujian	Fuzhou	Xiamen	Putian	Quanzhou	Ningde
Shandong	Jinan	Qingdao	Dongying	Yantai	Weifang
	Weihai	Rizhao	Binzhou		
Hubei	Wuhan				
Guangdong	Guangzhou	Zhuhai	Shenzhen	Shantou	Foshan
	Jiangmen	Zhanjiang	Maoming	Huizhou	Dongguan
	Jieyang				
Guangxi Zhuang Autonomous Region	Nanning	Beihai	Qinzhou		
Hainan	Haikou	Sanya			
Shanxi	Xi'an				
Gansu	Lanzhou				

Appendix 10: List of Marine-related Higher Institutions (Including Marine-related coefficient of proportionality)

1. Higher Institutions Directly under the Ministry of Education

Beijing University (0.0932) (the marine-related coefficient of proportionality is determined according to the proportion of marine-related majors in the total number of majors; the same approach is employed below),Tsinghua University (0.0256), Beijing Normal University(0.1373), China University of Geosciences (Beijing) (0.2381), Tianjin University (0.0256), Dalian University of Technology (0.0886), Shanghai Jiao Tong University(0.0484), Nanjing University(0.1163), Hehai (River Sea) University(0.9020), Zhejiang University(0.1102), Xiamen University(0.0707), Ocean University of China(0.8462), Wuhan University(0.0645), China University of Geosciences(Wuhan)(0.2258), Zhongshan University(0.1280), Tongji University(0.0859), East China Normal University(0.0789), Huazhong (Central China) University of Science and Technology(0.0566), South China University of Technology(0.0490).

2. Higher Institutions under the Ministry of Industry and Information Technology

Harbin Institute of Technology (0.0462)

3. Higher Institutions under the Ministry of Transportation

Dalian Maritime University (0.9348)

4. Local Higher Institutions

Shanghai Ocean University(0.3191), Guangdong Ocean University(0.2200), Dalian Ocean University(0.9545), Zhejiang Ocean University(0.8913), Ningbo University(0.1935), Jimei University(0.2388), Nanjing University of Information Science and Technology(0.2759).

Appendix 11: List of Marine-related Group of Disciplines (Discipline Classifications of the Ministry of Education)

Attached Table 11-1 List of Marine-related Disciplines (Discipline Classifications of the Ministry of Education)

Code	Disciplines	Descriptions
140	**Physics**	
14020	Acoustics	
1402050	Underwater Acoustics and Ocean Acoustics	Formerly known as "Underwater Acoustics".
1403064	Ocean Optics	
170	**Geoscience**	
17050	Geology	
1705077	Geology of Petroleum and Natural Gas	Including Geology of Natural Gas Hydrate.
17060	Marine Science	
1706010	Marine Physics	
1706015	Marine Chemistry	
1706020	Marine Geophysics	
1706025	Marine Meteorology	
1706030	Marine Geology	
1706035	Physical Oceanography	
1706040	Marine Biology	
1706045	Marine Geography &Estuarine and Coastal Science	Formerly known as "Estuarine and Coastal Science".
1706050	Marine Investigation and Monitoring	
	Marine Engineering	See 41630.
	Marine Surveying	See 42050.
1706061	Remote Sensing Oceanography	Also called "Satellite Oceanography".
1706065	Marine Ecology	
1706070	Environmental Oceanography	
1706075	Marine Resources Science	
1706080	Polar Science	

Code	Disciplines	Descriptions
1706099	Other Disciplines of Marine Science	
240	**Fishery Science**	
24010	Basic Disciplines of Fishery Science	
2401010	Aquatic Chemistry	
2401020	Geography of Fishery	
2401030	Aquatic Biology	
2401033	Aquatic Genetics and Breeding Science	
2401036	Aquatic Animal Medicine Science	
2401040	Aquatic Ecology	
2401099	Other Disciplines of the Basic disciplines of Fishery Science	
24015	Aquatic Multiplication Science	
24020	Aquaculture science	
24025	Aquatic Feed Science	
24030	Aquatic Protection Science	
24035	Fishing Science	
24040	Storage and Processing of Aquatic Products	
24045	Aquatic Engineering	
24050	Aquatic Resource Science	
24055	Aquatic Economics	
24099	Other Disciplines of Fishery Science	
340	**Military Medicine and Special Medicine**	
34020	Special Medicine	
3402020	Submarine Medicine	
3402030	Nautical Medicine	
413	**Engineering and Technology Related to Information and Systems Science**	
41330	Systemic Application of Information Technology	
4133030	Marine Information Technology	
416	**Engineering and Technology Related to Natural Science**	

Code	Disciplines	Descriptions
41630	Marine Engineering and Technology	The original code is 57050 and formerly known as "Marine Engineering".
4163010	Marine Engineering Structure and Construction	The original code is 5705010.
4163015	Seabed Mineral Development	The original code is 5705020.
4163020	Utilization of Seawater Resources	The original code is 5705030.
4163025	Marine Environmental Engineering	The original code is 5705040.
4163030	Coastal Engineering	
4163035	Offshore Engineering	
4163040	Deep Sea Engineering	
4163045	Marine Resources Development and Utilization Technology	Including Ocean Mineral Resources, Seawater Resources, Marine Biology, Marine Energy Development Technology, etc.
4163050	Ocean Observation and Forecasting Technology	Including Ocean Underwater Technology, Ocean Observation Technology, Ocean Remote Sensing Technology, Ocean Forecasting Technology, etc.
4163055	Marine Environmental Protection Technology	
4163099	Other Disciplines of Marine Engineering and Technology	The original code is 5705099.
420	**Science and Technology of Surveying and Mapping**	
42050	Marine Surveying and Mapping	
4205010	Marine Geodetic Survey	
4205015	Marine Gravity Survey	
4205020	Marine Magnetic Survey	
4205025	Ocean Spring Layer Survey	
4205030	Ocean Sound Velocity Survey	
4205035	Hydrographic Survey	
4205040	Seafloor Topography Survey	
4205045	Hydrographic Charting	
4205050	Marine Engineering Survey	
4205099	Other Disciplines of Marine Surveying and Mapping	

Code	Disciplines	Descriptions
480	**Energy Science and Technology**	
48060	Primary Energy	
4806020	Petroleum and Natural Gas Energy	
4806030	Hydro energy	Including Ocean Energy, etc.
4806040	Wind Energy	
4806085	Natural Gas Hydrate Energy	
490	**Nuclear Science and Technology**	
49050	Nuclear Power Engineering Technology	
4905010	Marine Nuclear Power	
570	**Hydraulic Engineering**	
57010	Basic Disciplines of Hydraulic Engineering	
5701020	Rivers and Coastal Dynamics	
580	**Transportation Engineering**	
58040	Waterway Transport	
5804010	Navigation Technology and Equipment Engineering	Formerly known as "Nautical Navigation".
5804020	Ships Communication and Navigation Engineering	Formerly known "Navigation Building & Navigation Mark Engineering".
5804030	Waterway Engineering	
5804040	Harbour Engineering	
5804080	Marine Technology and Equipment Engineering	
58050	Ships and Warships Engineering	
610	**Environmental Science and Technology & Resource Science and Technology**	
61020	Environmental Science	
6102020	Water Environment Science	Including Marine Environmental Science.
620	**Safety Science and Technology**	
62010	Basic Disciplines of Safety Science and Technology	
6201030	Catastrophology	Including Disaster Physics, Disaster Chemistry, Disaster Toxicology, etc.

Code	Disciplines	Descriptions
780	**Archaeology**	
78060	Specialized Archaeology	
7806070	Underwater Archaeology	
790	**Economics**	
79049	Resource Economics	
7904910	Marine Resource Economics	
830	**Military Science**	
83030	Science of Campaigns	
8303020	Science of Naval Campaigns	
83035	Science of Tactics	
8303530	Science of Naval Tactics	

Notes:

According to percentage of the number of marine-related disciplines (third level disciplines) that are included in the second level disciplines in the total number of third level disciplines, the marine-related coefficient of proportionality of the second level disciplines are determined as follows: acoustics (0.06), optics (0.06), geology (0.04), marine science (1), basic disciplines of fishery science (1), aquatic multiplication science(1), aquaculture science (1), aquatic feed science (1), aquatic protection science (1), fishing science (1), storage and processing of aquatic products (1), aquatic engineering (1), aquatic resource science (1), aquatic economics (1), other disciplines of fishery science (1), special medicine (0.33), systemic application of information technology(0.25), marine engineering and technology (1), marine surveying and mapping (1), primary energy (0.36), nuclear power engineering technology (0.20), basic disciplines of hydraulic engineering (0.25), waterway transport (0.56), ships and warships engineering (1), environmental science (0.17), basic disciplines of safety science and technology (0.17), specialized archaeology (0.11), resource economics (0.17), science of campaigns (0.17), science of tactics (0.17).

Compilation Explanation

Commissioned by the Department of Science and Technology, State Oceanic Administration, the First Institute of Oceanography of SOA has been embarking on the measurement and calculation of marine innovation indicators as from 2006 and officially initiated the research on national marine innovation index in 2013 in order to respond to national marine innovation strategies, and to offer service to the construction of national innovation system. "*Evaluation Report of National Marine Innovation Index 2016*" is the fourth issue of related series of evaluation reports. The relevant information is explained as follows:

1. Needs Analysis

Innovation-driven development has become a national development strategy of our country. "*The Decision of the Central Committee of the Communist Party of China on Major Issues Concerning Comprehensively Deepening Reforms*" explicitly put forward "building national innovation systems". Marine innovation is a key field in establishing innovation-oriented country and an important component of national innovation system. To explore the building of national marine innovation index and evaluate the national marine innovation capability of our country are of great significance to the construction of a marine power. The necessity of compiling "*Evaluation Report of National Marine Innovation Index 2016*" is mostly manifested in the following four aspects:

(a) Urgent demand to find out the situation of marine innovation in our country.

To collect marine innovation data such as marine economic statistics, science and technology statistics and S&T achievements registration, so as to get a clear picture of the marine innovation situation of our country is the foundation of objective analysis of marine innovation development trends of our country.

(b) Objective demand to grasp the development trend of marine innovation in our country.

To mine and analyze marine innovation data from the four aspects i.e. marine innovation environment, marine innovation resources, marine knowledge creation and marine innovation performance so as to gain a deeper knowledge of marine innovation and development trends in our country is the essential prerequisite of understanding the marine innovation path and mode.

(c) Actual demand to calculate accurately the key indicators of marine innovation in our country.

To calculate and predict marine innovation key indicators such as the contribution rate of marine S&T progress, and the transformation rate of marine S&T achievements, etc. can give a realistic picture of the quality and efficiency of marine innovation in our country, providing a series of important indicators for the formulation of marine innovation policy.

(d) Current demand to gain a comprehensive understanding of international marine innovation development.

To analyze the development trends of international marine innovation on both the scientific research level from the perspectives of the institutes that publish marine scientific papers, and the influence of those papers, and the technological R&D level from the application agencies of marine patents, technological layout and protection strategies, etc. can help us comprehensively understand the trend of international marine developments, providing references for the development of marine innovation in our country .

2. Compilation Basis

(a) The Report of the Eighteenth National Congress of the Communist Party of China (CPC)

With "becoming an innovative country" as a goal of building a moderately prosperous society in all aspects and deepening reform and opening up in an all-round way, the "*report*" put forward "the implementation of the innovation-driven development strategy" and points out that "S&T innovation provides strategic support for raising productivity forces and boosting overall national strength" and "we should efficiently allocate and fully integrate innovation resources, and ensure that the wisdom and strength of the whole society are directed toward promoting innovation–driven development".

(b) The Report of the Fifth Plenary Session of the Eighteen Central Committee of the Communist Party of China

The "*report*" put forward that "we must give innovation top priority in overall national development, continue to make innovations in theory, institutions, science and technology, culture and other fields, implement innovation throughout all the work of the Party and the state and make innovation become common practice in the whole of society".

(c) Outline of the National Innovation-driven Development Strategy

The "*outline*" which was issued by the State Council of the CPC Central Committee in May, 2016, pointed out that "the implementation of the innovation-driven development strategy and the emphasis of the importance of S&T innovation as the strategic support for raising productivity forces and comprehensive national strength put forward by the Eighteenth National Congress of the Communist Party of China (CPC) must be the focus of national development, which is the important development strategy determined by the CPC Central Committee at new development stage to base on the overall situation, face the world, focus on the key points and promote the development of the whole country."

(d) Outline of the '*13th Five-Year Plan*' for National Economic and Social Development of People's Republic of China

The "*outline*" put forward the major innovation-driven indicators, strengthened the leading role of

S&T innovation and pointed out that "development must be based on innovation with S&T innovation as focus and talents development as support to promote the organic combination of S&T innovation with mass entrepreneurship and innovation, and build more leading development models driven by innovation and giving more play of first mover advantage".

(e) Vision and Actions on Jointly Building Silk Road Economic Belt and 21st-Century Maritime Silk Road

The "*Vision and Actions on Jointly Building Silk Road Economic Belt and 21st-Century Maritime Silk Road*" put forward new development ideas to "create new systems and mechanisms of open economy, step up scientific and technological innovation, develop new advantages for participating in and leading international cooperation and competition, and become the pace-setter and main force in the 'Belt and Road Initiative', particularly the building of the 21st-Century Maritime Silk Road".

(f) The Decision of the Central Committee of the Communist Party of China on Major Issues Concerning Comprehensively Deepening Reforms

The "*decision*" explicitly stated "the building of national innovation system".

(g) National "*13th Five-Year*" Planning for Scientific and Technological Innovation

The "*planning*" put forward that "the period of the '*13th five-year*' is the decisive stage of building a moderately prosperous society in all aspects and becoming one of the innovative countries. It is also the key stage of implementing innovation-driven development strategies deeply and deepening the reform of scientific and technological system across the board. Facing the world and based on the whole situation, we must conscientiously carry out the decisions of the Central Committee of the Party and the State Council, thoroughly understand and accurately grasp the new requirements in the context of new normal of economic growth and new trend of S&T innovation both at home and abroad, explore the new path of innovation development systematically, open up new realm of development led by S&T innovation, stride forward the ranks of innovative countries and speed up the construction of world-class S&T power".

(h) Overall Planning of Marine Scientific and Technological Innovation

At the first working conference on strategic research, the "*Overall Planning of Marine Scientific and Technological Innovation*" urged that efforts be made "to conduct the research of marine strategy on the basis of '*overall planning' and 'innovation*'", "to recognize the innovation path and modes and to take stock of '*overall resources*'".

(i) "*13th Five-Year*" Special Plan for Scientific and Technological Innovation in Marine Field

The "*special plan*" stated clearly that "the period of the '*13th five-year*' is the decisive stage of building a moderately prosperous society in all aspects and the key stage of implementing innovation-driven development strategies and building a maritime power", "national marine S&T system should be further established and improved, marine S&T innovation capability should be enhanced and the supporting role of S&T innovation to the improvement of marine industry development should be strengthened significantly".

(j) Outline of the National "*12th Five Year Plan*" for Marine Science and Technology Development

The "*outline*" explicitly stated that during the "*12th Five-Year Pla*n" period, the overall goals of marine science and technology development are to "markedly boost the capability of independent innovation", "significantly enhance the technological innovation capability in coastal areas, build a more sound scientific and technological marine innovation system, and realize a contribution rate of up to 60% of marine science and technology to marine economy, thereby basically forming the sustainable development capability of the marine economy and the marine cause driven by marine science and technology innovation"

(k) Outline of the National Marine Economy Development Plan

The "*outline*" proposed to "gradually build China into a marine power".

(l) Outline of Marine Development by means of Science and Technology (2008-2015)

The "*outline*" stated clearly that "we must guide and promote the transformation and industrialization of marine S&T achievements, accelerate the development of marine industries, support and promote faster and better development of the marine economy in coastal areas".

(m) National "*12th Five-Year*" Development Plan for Marine Industry

It pointed out that "marine industry must be placed in a very important strategic position" and "we must step up the development of marine industry and strive to build our country into a marine power".

(n) The Development Plan of the Science and Technology Support System for the Construction of Maritime Power

According to the "*plan*", the basic objectives of a S&T support system for the construction of a marine power are to "optimize the overall layout of marine scientific and technological development and marine development by means of science and technology based on the general requirements of the development plan for marine industry, rely on the existing scientific research strength and S&T resources with innovative advantage of the central and local governments", and "establish a marine S&T support system to serve the goals of building our country into a marine power and strengthen the capability of scientific marine undertakings and enhance the core competitiveness of marine industry".

(o) National Medium and Long-Term Program for Science and Technology Development (2006-2020)

The "*National Medium- and Long-Term Program for Science and Technology Development (2006-2020)*" put forward that "we must place the strengthening of indigenous innovative capability at the core of economic restructuring, growth model change, and national competitiveness enhancement. Building an innovation-oriented country is therefore a major strategic choice for China's future development". It also pointed out that the guiding principles for our S&T undertakings were 'indigenous innovation, leapfrogging in key fields, supporting development, and leading the future'", and emphasizes that "we should greatly advance the construction of national innovation system with Chinese characteristics and dramatically enhance the nation's indigenous innovation capability".

(p) National "*12th Five-Year*" Development Plan of Science and Technology

The "National '*12th Five-Year*'" Development Plan of Science and Technology pointed out that "'the *12th Five-Year Plan*' is the key period of building a moderately prosperous society in all aspects in China, and a crucial stage of improving the capability of independent innovation and building the innovative country" and "we must give full play to the important supporting role of S&T progress and innovation to accelerate the transformation of economic development mode".

(q) National "*13th Five-Year*" Special Plan for Scientific and Technological Innovation

The "*plan*" pointed out that innovation was the first driving force to lead development. The "*plan*" has made deployments from 6 aspects for S&T innovation in order to deeply implement the innovation-driven development strategies and support the supply side structural reform. The "*plan*" put forward that by 2020, the world ranking of China 's National Comprehensive Innovation Capability will have moved up to No.15 from current No.18; the contribution rate of S&T progress will be raised up to 60% from current 55.3%; R&D input strength will be increased up to 2.5% from current 2.1%.

3. Data Sources

The data used by "*Evaluation Report of National Marine Innovation Index 2016*" are from the following sources:

(a) China Statistical Yearbook

(b) China Marine Statistical Yearbook

(c) Statistical Data of the Ministry of Science and Technology

(d) Statistical Data of Marine-related Higher Institutions and Marine-related Disciplines of the Ministry of Education

(e) Data on Marine Science Papers and Marine Patents from Lanzhou Centre for Documentation and Information of the Chinese Academy of Social Sciences

(f) Database of Chinese Science Citation Database, CSCD

(g) Database of Science Citation Index Expanded, SCIE

(h) Related Data of Regional Demonstration of National Marine Innovation Development

(i) Registration Data on Marine S&T Achievements

(j) Compilation of Science and Technology Statistical Data in Higher Institutions

(k) Other publications

4. Compilation Process

Commissioned by the Department of Science and Technology, Marine Policy Research Center of the First Institute of Oceanography of SOA organized the compilation of "*Evaluation Report on National Marine Innovation Index 2016*". Lanzhou Documentation and Information Center of the Chinese Academy of Sciences co-wrote the marine papers and patents, analyzed the situations of international marine science and technology research and other parts; Qingdao National Laboratory of Marine Science and Technology did special analysis to marine national laboratory; the Department of Science and Technology of SOA provided the relevant content of the regional demonstration of marine economic innovation and development in China; Department of Innovation and Development of the Ministry of Science and Technology, Science and Technology Department of the Ministry of Education, National Marine Data and Information Service, School of Management, Huazhong (Central China) University of Science and Technology and other units and departments provided data support. The specific compilation process is divided into three stages, which are the preparation stage, data calculation and report preparation stage, consultation and revision stage respectively. Specific introductions are as follows:

(a) Preparation stage

Formation of basic ideas from January, 2017 to February, 2017. The first, second and third issues of series of reports on national marine innovation index assessment, namely the "*Tentative Evaluation Report of National Marine Innovation Index 2013*", "*Tentative Evaluation Report of National Marine Innovation Index 2014*", and "*Evaluation Report of National Marine Innovation Index 2015*" were published in May, 2015, December, 2015 and December, 2016 respectively. In early 2017, the compiling idea on the "*Evaluation Report of National Marine Innovation Index 2016*" was formed and a specific compiling plan of it was made and submitted to the Department of Science and Technology of the State Oceanic Administration on the basis of the previous work of the "*Evaluation Report of National Marine Innovation Index 2015*" after many rounds of research and communication to summarize experiences and find out deficiencies of the previous three issues.

Data collection. In January, 2017, we smoothly obtained S&T innovation data of marine scientific research institutes, related data of *Compilation of Science and Technology Statistical Data in Higher Institutions* and marine-related S&T innovation data extracted by marine-related higher education institutions based on marine-related disciplines (first level) from Science and Technology Statistics Information Center of Huazhong (Central China) University of Science and Technology and Science and

Technology Department of the Ministry of Education. Meanwhile, working with Lanzhou Centre for Documentation and Information of the Chinese Academy of Social Sciences, we collected data of SCI papers in marine fields and marine patents, and so on.

Building of the report compiling team and the indicator calculation team. In January, 2017, under the guidance of the Department of Science and Technology of the State Oceanic Administration, and the assessment advisory group of national marine innovation index, and on the basis of the original compiling teams of the "*Evaluation Report of National Marine Innovation Index 2015*", we built the compiling and indicator calculation teams of the "*Evaluation Report of National Marine Innovation Index 2016*", which are composed of staff from Marine Policy Research Center, the First Institute of Oceanography of the State Oceanic Administration, China, and Lanzhou Centre for Documentation and Information of the Chinese Academy of Social Sciences.

(b) Data calculation and report preparation stage

Data processing and analysis. From January to February, 2017, we have processed and analyzed the S&T innovation data of marine scientific research institutes and data from "*China Statistical Yearbook*", "*China Marine Statistical Yearbook*", "*Compilation of Science and Technology Statistical Data in Institutions of Higher Education*", and marine-related S&T innovation data extracted by marine-related higher education institutions based on marine-related disciplines (first level).

Data calculation. From the 20th of February to the 20th of March, 2017, the calculation team has calculated the contribution rate of marine S&T progress and transformation rate of marine S&T achievements. National marine innovation index and regional marine innovation index have also been calculated according to the corresponding evaluation methods.

Compilation of the first draft of the report. From the 21st of March to the 20th of April, 2017, the first draft of the report was completed based on the results of data analysis and indicator calculation.

The first round of data review. From the 21st of April to the 7th of May, 2017, the calculation team has conducted the first round of data review, focusing on data sources, data processing procedures, figures and tables.

Revision of the second draft of the report. From the 8th to the 22nd of may, 2017, the first draft of the report has been revised according to the results of data review and the calculation results of indicators, and the second draft was completed.

The second round of data recheck. From the 23rd to the 31st of June, 2017, the calculation team has taken the reverse review method and conducted the second round of data review, reviewing figures, tables, data processing and data sources one by one according to the report content.

Solicitation of opinions on a small-scale. From the 1st to the 10th of June, 2017, internal opinions have been solicited on a small scale.

Improvement of the third draft of the report. From the 11th to 20th of June, 2017, the second draft of the report has been improved according to the results of the second round of data review and opinions solicited on a small scale, and the third draft was completed.

(c) Stages of report review, revision and improvement

Preliminary review of administration departments. From the 21st to the 27th of July, 2017, the report was submitted to the Department of Science and Technology of the State Oceanic Administration for review and revised according to the suggestions of that review. The revision focus is Chapter VI, regional demonstration of marine economic innovation development, and related data were supplemented.

Internal review and revision of the fourth draft of the report. From the 28th of June to the 5th of July, 2017, we have organized internal review and revised the report according to the suggestions of the internal review.

Recheck of calculation process. From the 6th to the 14th of July, 2017, the calculation team has rechecked the calculation process, focusing on formulas of calculation, parameters and accuracy of results. The report has been improved and perfected according to the results of the recheck and the fourth draft was completed.

The second review of administration departments. From the 15th to the 28th of July, 2017, the report was submitted to the Department of Science and Technology, SOA for review and revised according to the review suggestions.

Review of advisory panel. From the 29th of July to the 22nd of August, 2017, advisors have been organized to review the report. The report has been revised according to the suggestions of the advisory panel.

Panel discussion. On August 23, 2017, experts were invited and organized to make comments by national marine information center and the report was revised based on the feedbacks of the experts.

Pre-editing of the publishing house. In August, 2017, the paper version of the report was submitted to the Editorial Department of Ocean Publishing House for pre-editing.

Final revision and formation of the official version. In September, 2017, the official version was completed.

5. Adoption of Opinions And Suggestions

More than 70 person times have been solicited for opinions and over 900 pieces of opinions and suggestions have been received after summary.

More than 620 pieces of opinions and suggestions have been adopted, with the adoption rate of feedback and suggestions being approximately 69%.

Instructions on Updates

1. Some chapters and contents are added and deleted

(a) Chapter VII "Specific Analysis on the Input-Output Efficiency of Marine Science and Technology in China" is newly added.

(b) The first section of Chapter II "The age structure of S&T personnel" is deleted due to the quality problem of data.

(c) Chapter VII "Specific Analysis on Marine Innovation Capability of Chinese Enterprises" and Chapter VIII "Specific Analysis on the Layout of Marine S&T Power of Chinese Cities" are deleted.

2. Both domestic and international data are updated

(a) The data of international marine-related innovation papers are updated. The original data are updated until 2015, used to analyze marine innovation output achievements and make comparative analysis of domestic and foreign marine innovation papers.

(b) The data of international patents are updated. The original data are updated until 2015, used to analyze marine innovation output achievements and make comparative analysis of domestic and foreign marine innovation patents.

(c) The data of marine-related higher institutions are updated. The original data are updated until 2015, used to analyze marine innovation development in higher institutions.

(d) The domestic data are updated. The original data used by the evaluation indicators of national marine innovation are updated until 2015 and the evaluation indicator of regional marine innovation sub-index are updated until 2015.

3. Some chapters and contents are revised and perfected

Methods description is revised and perfected. The methods description with regard to the output efficiency of marine input is revised and perfected, which gives a detailed introduction to data processing methods and evaluation calculation processes.